Wind Power for the Electric-Utility Industry

Arthur D. Little Books

A series of books on management and other scientific and technical subjects by senior professional staff members of Arthur D. Little, Inc., the international consulting and research organization. The series also includes selected nonproprietary case studies.

Acquisition and Corporate Development   James W. Bradley and Donald H. Korn

Bankruptcy Risk in Financial Depository Intermediaries: Assessing Regulatory Effects   Michael F. Koehn

Board Compass: What It Means to Be a Director in a Changing World
   Robert Kirk Mueller

Career Conflict: Management's Inelegant Dysfunction   Robert Kirk Mueller

The Corporate Development Process   Anthony J. Marolda

Corporate Responsibilities and Opportunities to 1990   Ellen T. Curtiss and
   Philip A. Untersee

The Dynamics of Industrial Location: Microeconometric Modeling for Policy Analysis   Kirkor Bozdogan and David Wheeler

Systems Methods for Socioeconomic and Environmental Impact Analysis
   Glenn R. De Souza

The Incompleat Board: The Unfolding of Corporate Governance
   Robert Kirk Mueller

Energy Policy and Forecasting   Glenn R. De Souza

Communications Network Analysis   Howard Cravis

Board Score: How to Judge Boardworthiness   Robert Kirk Mueller

International Telecommunications: User Requirements and Supplier Strategies
   Edited by Kathleen Landis Lancaster

New Dimensions of Project Management   Edited by Albert J. Kelley

Wind Power for the Electric-Utility Industry: Policy Incentives for Fuel Conservation   Frederic March, Edward H. Dlott, Donald H. Korn, Frederick R. Madio, Robert C. McArthur, and William A. Vachon

Wind Power for the Electric-Utility Industry

Policy Incentives for Fuel Conservation

Frederic March
Edward H. Dlott
Donald H. Korn
Frederick R. Madio
Robert C. McArthur
William A. Vachon

An Arthur D. Little Book

LexingtonBooks
D.C. Heath and Company
Lexington, Massachusetts
Toronto

333.79
W 763m

Library of Congress Cataloging in Publication Data

Main entry under title:

Wind power for the electric-utility industry.
   "An Arthur D. Little Book."

   Includes index.
   1. Electric utilities—Energy conservation—Government policy—United States.   2. Wind power. I. March, Frederic.
HD9685.U5W56   333.79   81-48267
ISBN 0-669-05321-x   AACR2

Copyright © 1982 by D.C. Heath and Company.

Sponsored by the National Science Foundation, Grant Number: PRA 8000488.
All rights reserved. No part of this publication may be reproduced or transmitted in any form or by any means, electronic or mechanical, including photocopy, recording, or any information storage or retrieval system, without permission in writing from the publisher.

Published simultaneously in Canada

Printed in the United States of America

International Standard Book Number: 0-669-05321-x

Library of Congress Catalog Card Number: 81-48267

Contents

	Figures	viii
	Tables	ix
	Acknowledgments	xi
Chapter 1	Introduction	1
Chapter 2	Review of Wind-Energy Technology and Applications	7
	Background on Solar-Electric Technology	7
	Federal R&D for Large Wind Turbines	9
	Private-Company Large Wind-Turbine Production Activities	14
	Overview of Major-Utility Wind Programs	18
	Third-Party Wind-Turbine Enterprise	20
	Foreign Wind-Energy Programs	21
	Off-Shore Wind-Energy Projects	25
	Ultimate Utility Wind-Energy Potential	28
Chapter 3	Methodology of Study	31
	Background and Literature Review	32
	Simulation	43
Chapter 4	Perspectives on the Electric-Utility Industry	55
	The Regulatory Environment for Electric Utilities	55
	Utility Perceptions of Risk and Reward	58
	The Financial Condition of the Electric-Utility Industry	63
	Current Issues in Utility Regulation	68
	Recent Legislation and Policies Affecting Utility Investment in Wind Power	77
	State Initiatives: The California Experience	84
Chapter 5	Constructing a Utility Case Study	93
	The Wind Resource	94
	The Wind Project	100

	Synthetic-Utility Characteristics	100
	Simulation of Utility-Operating Characteristics	109
	Financial and Regulatory Characteristics	114
	Conventional Engineering-Economic Analysis	118
Chapter 6	**Financial and Policy Analysis of Scenarios**	125
	Issues in the Case Study	125
	Selection of Policy Alternatives for Study	128
	Results of the Financial Model	132
	Identification of Effective Policies	144
	Overriding Public-Policy Considerations	146
Chapter 7	**Summary of Major Findings**	151
	Utility Investment in Wind-Power Systems	151
	Investment-Evaluation Criteria Used by Utilities	152
	The Effects of Public-Policy Incentives	153
	Index	155
	About the Authors	161

Figures

2-1	Mod-2 Wind Turbine	13
2-2	Prototype of Alcoa ALVAWT Model #1238229-500 kW	15
3-1	Typical Methodology for Solar-Electric Value Analysis	34
3-2	Overview of Simulation for Case Study	44
3-3	Average Hourly Wind Velocity versus Power (MOD-2 1000 MW Cluster)	46
4-1	Trends in Selected Indicators of Utility Financial Health	64
4-2	Market Action	65
5-1	Annual Average Wind Power in the Northeast	95
5-2	Seasonal Average Wind Power in Pennsylvania	99
5-3	Average Monthly Wind Speed, Synthetic-Utility Site	102
5-4	Simulation Results: Speed versus Power Output	104
5-5	Monthly Peak-Load Variation	105
5-6	Synthetic-Utility Expansion Plan	106
5-7	Diurnal Load versus Wind Velocity	107
5-8	Monthly Variation of Wind-Power Output versus Peak Load	113
5-9	Breakeven Cost Results	119

Tables

1–1	List of Organizations Contacted	3
2–1	Summary of Federally Funded Large Wind Turbines for Utility Applications	11
2–2	Cost Projections for Federally Funded Large Wind Turbines	14
2–3	Key, Privately Funded, Large Wind Turbines	16
2–4	Swedish Wind Turbines	24
2–5	Estimated Displacement of Oil Using Large WECS	26
3–1	Summary of WECS Value Analysis	37
3–2	Capacity Credit Estimates	39
3–3	Value of Storage Estimates	41
4–1	Changes in S&P's Electric-Utility Bond Ratings	64
4–2	U.S. Electric-Power Industry Annual Growth Rates	66
4–3	Selected Financial Statistics: Investor-Owned Electric Utilities	67
4–4	Dispersed Supply Additions: Targets Established by the California Energy Commission and Utility Plans	87
5–1	Classes of Wind-Power Density at 10 m and 50 m	96
5–2	Areal Distribution of Wind-Power Classes in the Northeast	97
5–3	Monthly and Annual Average Wind Speeds at Boone, North Carolina	101
5–4	Wind-Energy System for Simulation	103
5–5	Wind Cluster: Construction and Operation Schedule	103
5–6	Comparison of Synthetic-Utility Equipment Mix with Regional Equipment Mix in 1979	107
5–7	Generating-Fuel Cost, 1985	108
5–8	Cost of Fossil Fuels Delivered to Steam-Electric Plants	108

5-9	Regional Fuel Prices, July 1980	109
5-10	Regional Fuel Prices, November 1980	109
5-11	Typical Fuel Prices Paid by Utilities in Northeast Region in April 1980	110
5-12	Production-Cost Simulation Results: Savings	111
5-13	Net Savings per Wind-Generated MWH	112
5-14	Fuel Displaced by Wind (10^3 MWH)	112
5-15	Fuel Displaced by Wind (Fuel Units)	113
5-16	Computer Input to the Financial Simulation Program (Sample)	116
6-1	Comparisons of Measures of Utility Financial Condition with and without Wind Investment	133
6-2	Utility Financial Measures for Policy Scenarios	134

Acknowledgments

Dr. Jack Allison of the National Science Foundation (NSF) was program manager and helped in the definition of the research work. Dr. Frank Huband later became the NSF program manager and provided valuable discussions. Dr. Peter E. Glaser, vice-president, and Dr. David M. Boodman, vice-president, constituted an Arthur D. Little, Inc. (ADL), Review Committee, which provided management support. Dr. W. Scott Nainis and Dr. Bette M. Winer of the Operations Research Section at ADL consulted on various aspects of the methodology. Harvey Salgo, Esq., of the law firm of Adamo, Lee and Salgo provided assistance in research of legal issues. Mr. Brian Curry, Mr. Frank Sabatino, and Mr. Don Burbank of Northeast Utilities provided two workshop sessions to help define scenarios and to assist in the analysis of results.

Mr. Danny Dees, Mr. Jim McDermott, and Mr. William Marsh of the General Electric Company, Electric Utility Systems Division, provided the computer programs used herein, as well as a supportive interest in the work. Ms. Mary A. Quigley typed the entire draft report manuscript. Ms. Margo E. Galloway typed the revisions and prepared the report for submission to the National Science Foundation.

The views expressed in this book are those of the authors, who take full responsibility for any errors of commission or omission, and do not necessarily represent the views of NSF or any other parties mentioned in this book.

1 Introduction

For many years technological innovation in the electric-utility industry has been characterized by economies of scale, which have involved the generating facilities as well as the transmission networks. Real savings were obtained as utilities invested in new technology, and projects were successfully implemented. One effect was to assure that investor-owned utilities were viewed in the financial markets as stable, reliable, steady-growth enterprises—a reasonably safe repository for conservative investment portfolios. Another effect was a steady decline in the real price of electrical energy to the consumer, as investments in new plants with their economies of scale reduced unit-operating costs.

For a considerable period, a major factor in the long-term price stability of electrical energy has been the low cost of oil, particularly during the time when the country was a net exporter of oil. From 1960 to 1970, according to the *Statistical Abstract of the United States*, the cost of a barrel (bbl) of crude oil slowly increased from about $3.00/bbl to $3.30/bbl. In real terms this was a reduction of 17 percent in the cost of the commodity. Low prices were enjoyed until the dramatic oil embargo of 1973, by which time the United States was importing sufficient oil to be adversely affected. By 1975 the average price of crude oil in the United States reached $15/bbl, and it increased steadily until very recently. The price of coal has also increased in the new energy marketplace.

In 1973, the first year of the shift to high crude-oil prices, according to the Department of Energy's *Monthly Energy Reviews*, electric utilities paid 11 cents per gallon, and by 1980 55 cents per gallon—a fivefold increase. In January 1981 New England Utilities paid 76 cents per gallon for residual oil, and 88 cents per gallon for distillate oil.

This single factor—the changing proportion of fuel cost in electrical production—induced a major structural shift in the industry. While expansion plans had included investments into oil-based plants because of their ease of operations compared to coal, their flexibility for load following as compared with nuclear, and the ability to more easily satisfy environmental constraints, a major shift occurred in which nuclear and coal plants rapidly began to dominate the new base-load planning horizon.

Meanwhile the effects of economy of scale began to diminish as nuclear and coal plants approached the 1000-MW range. While this issue is complex and somewhat debatable, it is clear that institutional factors related to environmental regulations and procedures involving lengthy public participation and intervenor groups in regulatory proceedings added to the costs of construction and increased the time from commitment of capital funds to actual energy production.

Several technological approaches have begun to emerge as promising in this kind of electric-utility environment, all of which are intended to reduce the high cost of fossil fuel BTU's consumed per net kilowatt-hour of electricity produced, and some of which may eventually reduce the need for conventional future baseload. (See chapter 4 for a discussion of new technological targets for the electric-utility industry in California.)

This book is based on a study performed under a grant from the National Science Foundation by an Arthur D. Little, Inc., research team, to demonstrate a systematic method for evaluating the economics of solar-electric/conservation technologies as fuel-savings investments for electric utilities, under alternative federal incentive policies. One of the findings is that, during the study period 1980–1981, the financial regulatory environment of the investor-owned electric utilities provided little or no incentive to invest in fuel-conserving technologies. Even if a project has a favorable benefit–cost ratio, utilities are not likely to make the investment unless the regulatory outlook is more favorable. The effects of various government-incentive policies on this situation can be understood with the aid of computer-based simulation analysis. The industry can probably be motivated to invest in fuel-conserving technology through specific policies at the federal and state levels that will improve the financial performance of the utility.

To develop and illustrate this thesis, we selected wind power as a near-term fuel-conservation technology that can significantly reduce fuel consumption in the electric-utility industry within the 1980s. The capabilities of wind-energy electric systems (WECS) for electric-utility grid applications are documented in chapter 2. Six major American manufacturers of large wind turbines are identified, and the major existing and currently planned WECS installations are described.

In order to provide a sound basis for exploring policy issues, we had many discussions (table 1–1) and searched the literature for previous policy and planning studies involving wind power. We found that none of the earlier studies treated the financial impact of the project on the utility. Therefore, we developed a methodology for this purpose, using existing computer models. This methodology is described in chapter 3.

When dealing with economic and financial issues in a complex regulated environment, like that of the electric-utility industry, it is essential to understand some of the key issues and controversies that have characterized planning and decision making. If the policymaker's aim is to improve the environment to stimulate investments into wind power, or other fuel-conservation technology, he or she has no choice but to enter the labyrinth of laws, rules, regulations, procedures, practices, and precedents that structure the relationships among the utility industry, the state public-utility commis-

Introduction

**Table 1-1
List of Organizations Contacted**

Utilities
Northeast Utilities, Hartford, Connecticut
New England Electric Systems, Westboro, Massachusetts
Dallas Power and Light Company, Dallas, Texas
Pacific Gas and Electric Company, San Francisco, California

State agencies
California Public Utility Commission
California Energy Commission
Massachusetts Public Utilities Commission

Federal organizations
Federal Energy Regulatory Commission
House Committee on Science and Technology
Department of Energy
Solar Energy Research Institute
Office of Technology Assessment
Council on Environmental Quality
Western Area Power Authority
National Science Foundation
Library of Congress

Other organizations
Edison Electric Institute
Electric Power Research Institute
American Public Power Association
American Wind Energy Association
New England Congressional Institute
Environmental Defense Fund, Berkeley, California
U.S. Windpower, Inc., Burlington, Massachusetts
Decision Focus, Inc., Palo Alto, California
The Synectic Group, Washington, D.C.
National Association of Regulatory Utility Commissioners

sions, and others who contribute to decisions on what is built, who pays how much, and what the investor may earn. Chapter 4 is a concise guide to these issues.

In order to construct a quantitative case study that would allow us to measure the effects of policy, we took the following steps:

We selected the northeast United States so that we could use actual regional wind resources.

We selected a specific wind machine and designed a project involving 1000 MW of installed wind capacity over a seven-year period.

A "synthetic" utility, specified by the Electric Power Research Institute, was selected as a surrogate for an actual utility typical of the northeast United States, and of the Middle Atlantic region in particular. The surrogate consists of a complete specification of existing and planned equipment, operating costs, financial history, and other data.

An engineering–economic analysis of the wind project was performed, showing the conditions under which the project may have a favorable outcome. This analysis involved detailed simulation of the interaction between the wind machines and the utility on an hourly basis, for each of several years.

The investments into the wind project, and the reduced fuel costs (as measured by the simulation) were used to modify the inputs into a realistic corporate financial model of the utility. This allowed us to measure the impact of the project on earnings, stock issued, bond coverage, and various other indicators of overall corporate financial health.

Chapter 5 shows how we accomplished the engineering–economic analysis, and chapter 6 reveals the financial analysis. To this point, we discovered that the major adverse effects of the wind project were limited to the financial performance in the initial five years—in subsequent years the overall financial health of the utility is improved by a sound project. The final analysis included the following:

A range of federal and state policies was selected in light of the background in chapter 4 and in consultation with representatives of the electric-utility industry. These policies are considered realistic and capable of being implemented.

The financial-analysis model was rerun for each policy to be tested, for each of fifteen future years.

The results were tabulated and consolidated to allow for rapid inspection, evaluation, and interpretation.

The findings were formulated and summarized (chapter 7).

While this work provides extremely valuable insights into the effects of policy on a typical fuel-conservation investment, there are no easy conclusions. Different state public-utility commissions would probably use these results to support different policies. The same may be true of the successive federal administrations. What clearly emerges, however, are the following:

The electric-utility industry, as currently regulated, is not motivated to invest in capital-intensive fuel-conservation technologies.

The major barrier to such investments, assuming technical feasibility, is the financial condition of the utility coupled with an uncertain regulatory environment in which utility management and stockholders perceive a high penalty for making a technological "bet" which doesn't work out as expected.

Therefore, new policies that address the financial barrier are needed if the industry is to be moved to actually assert technological leadership and innovate the use of resources such as wind power.

This book demonstrates that the primary effects of such policies can be quantified and measured at a reasonable cost, using existing computer models. The results provide valuable guidance and insights to policymakers and those who influence policy, who want to stimulate higher productivity at lower fuel consumption in generating and distributing electricity.

2 Review of Wind-Energy Technology and Applications

Background on Solar-Electric Technology

There are currently three technologies for solar-electric-power generation that hold near-term promise: (1) solar-thermal electric; (2) photovoltaic; and (3) wind. Longer-term technologies include ocean-thermal-electric conversion (OTEC) and the solar-power satellite. These are beyond the planning horizons of most electric utilities today.

Solar-thermal-electric systems use solar energy to heat a working fluid that drives a Rankine or Stirling cycle engine. Photovoltaic cells are solid-state devices that produce direct current when exposed to sunlight. Wind turbines use solar-energy indirectly, as solar heating of the earth's surface motivates air movements.

The three near-term technologies have in common the property of intermittency—that is, they produce a variable and uncontrolled quantity of energy. In all cases their value would be enhanced by the presence of storage. In the absence of system storage such as pumped hydro, utility planners today evaluate solar-electric technology in terms of the opportunity to conserve fuel when possible. They do not credit these technologies as contributing to firm baseload nor to meeting peak loads upon demand.

Briefly, utilities have other renewable-energy options available. First, hydroelectric power, like wind a form of indirect solar energy, has been a traditional technology, with and without storage (run of the river). Today utilities are reevaluating their hydropower potential by upgrading existing plants and seeking additional sites of smaller scale (low-head hydro). This too has the property of intermittency. Second, geothermal energy constitutes an inexhaustible source for reliable baseload generation and use upon demand. These technologies are of course well within the planning horizons of utilities with access to the resource. Third, biomass, tidal, and wave-energy conversion are being studied and may have significant future application. Finally, cogeneration conserves energy through systemwide efficiencies when commercial and industrial facilities generate electricity on site and use the heat discharged in the process to displace other fuels.

All of the preceding technologies can be evaluated for economic integration into utilities, using the methodology presented in chapter 3. In this study we wanted to focus on a new, near-term technology that offers an intermittent source of energy and that will be applied without storage. Thus in this chapter we briefly review solar-thermal electric, photovoltaic, and wind.

We concluded that wind energy is the near-term intermittent technology with the greatest potential for large-scale power generation in American utilities. Thus in the following sections we provide a more detailed review of the status of wind-energy technology and the precedent for its use to date. Since wind-, solar-thermal, and photovoltaic-energy technologies are undergoing rapid development, and new facilities are being planned, this review can be considered only a more or less current snapshot, subject to rapid change in the next few years.

Solar-Thermal Electricity Conversion

The solar-thermal electricity conversion (STEC) technology with most promise for utility scale application is the "power tower" for which several acres of remote controlled mirrors (heliostats) reflect the sunlight onto an absorbtive boiler mounted on a tall tower. Steam produced by the boiler is supplied to turbine generators of conventional design. Although only in the prototype stage at present, the technology has performed well enough in field tests to encourage projections of megawatt-size systems for commercial application by the 1990s [1]. STEC is a leading recipient of research funds for solar generation of electricity. Stobaugh and Yergin [2] point out that the solar-thermal program commands nearly 20 percent of the entire federal budget for solar energy, and this trend is expected to continue through the 1980s.

The cost of the heliostats with their high-precision focused mirrors is the major economic barrier facing power-tower technology, accounting for two-thirds of the plant costs. The U.S. Department of Energy (DOE) hopes to reduce costs tenfold, to $70/m^2$, but leaders in industry see $140/m^2$ as a more realistic goal [2], entailing projected total plant costs of $2000 per kilowatt by the year 2000.

STEC fits well with the established utility infrastructure of producing, marketing and distributing energy. If the prices of conventional fuels experience rapid increases and the technology is proven, solar-thermal power generation will become increasingly attractive, and current economic and technical barriers will diminish in importance.

Photovoltaic Electricity Generation

Photovoltaic (PV) cells ("solar cells") are semiconductor devices that convert sunlight directly into electricity. They have no moving parts and do not employ working fluids. They are also easy to operate and extremely reliable. Another attraction is that silicon, the principal raw material present-

Wind-Energy Technology and Applications

ly used in one PV manufacturing process, is the most abundant solid element on earth.

Sandia Laboratories have identified the present high cost of the PV devices as the largest barrier to their widespread use [3].[1] Development work is needed to reduce the cost of manufacturing the single-crystal silicon cells currently used or to increase the efficiency of potentially cheaper amorphous silicon and other thin-film solar cells. Sandia [3] reports that it is technically possible to reduce PV cell costs (1980 dollars) to $2.80 per peak watt (producing electicity at $.20–$.70 kWh) during the next two years. At that price PV systems will likely become cost effective for remote and/or Third World applications. When cell costs drop to $.15–$.50 per peak watt, central stations can be expected to supply electricity at $.06–$.09 per kWh according to DOE. The time frame 1990–2000 is the earliest that this can be expected to happen.

Wind-Energy Conversion Systems (WECS)

WECS are the leading technology for near-term utility applications of intermittent solar technologies. The intensive development of WECS is described in the following sections that provide a description of the federal R&D efforts, private wind-turbine production, foreign wind energy programs, and other aspects. This demonstrates that WECS now constitute a genuine generating option to electric utilities. We expect an enlargement of private-sector activities as early sales to electric utilities demonstrate the commercialization potential.

Federal R&D for Large Wind Turbines

The key technical characteristics of the federally funded large wind turbines are summarized in table 2–1. These include horizontal-axis wind turbines (HAWTs) developed with funds managed by the NASA Lewis Research Center (LeRC) and the vertical-axis Darrieus (VAWTs) being developed under programs managed by the Sandia Laboratories. Each HAWT machine design shown in table 2–1 is based on the knowledge gained from testing previous machines along with the outputs of numerous supporting research and technology (SR&T) efforts at LeRC. The first generation WTs (MOD-0, MOD-OA, and MOD-1) are basically research tools designed to gain engineering and operational experiences with machines of various sizes under different operating loads, environments, and in different types of utility networks. Second-generation WTs (MOD-2, and so on) employ a series of advanced concepts that are expected to increase machine reliability and reduce the cost of energy (COE) produced. Some of the major machines that have been developed are described in the following paragraphs.

MOD-O. The MOD-O WT is located at NASA's Plumbrook facility near Sandusky, Ohio. Since 1975 it has been used as an engineering test bed for evaluating advanced-design concepts and validating the analytical methods and computer codes used to design later-generation machines. As such, the MOD-O operates only in support of engineering tests.

MOD-OA. The MOD-OA is essentially the same design as the MOD-O machine except for a larger generator (200 kW) larger gearbox, and slightly modified blades. The MOD-OA project has been intended to demonstrate the technical feasibility of wind turbines in four different utility applications (see table 2–1). They are essentially engineering test machines, providing information on long-term performance, component reliability, and maintenance requirements of utility-size machines. The four sites chosen for the WTs are providing useful data on machine behavior in different types of utilities, wind regimes, and environmental conditions.

MOD-1. In design philosophy the MOD-1 WT is also a scaled-up version of the MOD-O WT. The main goal of the MOD-1 WT is to obtain early operating experience with a megawatt-scale machine. This is a two-bladed, 61-m (200 ft) diameter machine with a rated power of 2 MW (reduced to 1.5 MW in later 1980). The blades are of welded steel, and the rotor is located downwind of the tower. Full-span pitch control is used to hold the rotor speed constant at 23 rpm (formerly 35 rpm). The tower is steel, tubular truss design; the overall design of the tower is predicated on the need to maintain high natural vibration frequencies in the system (for example, frequencies greater than twice the blade-rotation frequency) to avoid high resonant dynamic responses. This approach resulted in a heavy, rigid tower configuration with attendant high-system costs. Much of the experience gained from the MOD-1 program has been applicable to the MOD-2 effort. In January 1981, the MOD-1 machine sustained a failure of the low-speed shaft assembly due to fatigue loads. Funding constraints restricted the repair and restart of the machine. Present plans call for the MOD-1 to be decommissioned.

MOD-2. The MOD-2 WT incorporates many advanced-design features for reducing machine weight and cost that were identified during the MOD-OA and MOD-1 programs. The machine is a two-bladed, 21.5-m (300 ft) diameter WT with a 2.5-MW rating. Partial-span pitch control is used (versus full-span control) to control rotor speed and power. A teetered rotor design allows the blades to move five degrees in and out of the vertical plane

Table 2-1
Summary of Federally Funded Large Wind Turbines for Utility Applications

Machine Characteristics	MOD-0	MOD-OA	MOD-1	MOD-2	ALCOA 100 kW
Machine type	Downwind HAWT	Downwind HAWT	Downwind HAWT	Upwind HAWT	Darrieus VAWT
Rated power, kW	100 kW	200 kW	1500 kW	2500 kW	100 kW
Rotor diameter: m (ft)	38.1 (125)	38.1 (125)	61 (200)	91.5 (300)	17 (55)
Hub height: m (ft)	30.5 (100)	30.5 (100)	42.7 (140)	61 (200)	15.1 (49.7)
Cut-in wind speed[a] m/s (mph)	4.0 (9.0)	5.8 (13.0)	7 (15.6)	6.3 (14)	5.4 (12)
Rated wind speed[a] m/s (mph)	6.5 (14.5)	9.7 (21.7)	14.6 (32.6)	12.3 (27.7)	13.4 (30)
Cut-out wind speed[a] m/s (mph)	15.3 (34.2)	17.9 (40.0)	19.0 (42.6)	20.1 (45)	26.9 (60)
Expected annual energy output (MWH)[b]	—	820	3700	9260	160
Machine locations	Sandusky, Ohio	Clayton, N. Mex. Culebra Island, P.R. Block Island, R.I. Oahu, Hawaii	Boone, N.C.	Goodnoe Hills, Wash. (each machine in 3-unit cluster)	Rocky Flats, Colo. Bushland, Tex. Martha's Vineyard, Mass.

Source: Compiled by Arthur D. Little, Inc.
[a] Wind speeds measured at Hub Height.
[b] Calculation based on a $\bar{V} = 6.3$ m/s (14 mph) site and a 90-percent Machine Availability Factor.

of rotation in order to reduce wind gust and shear loads on the blades and overall system. The gearbox is a compact, epicyclic design that is lighter in weight than the parallel-shaft gearbox used on the MOD-OA and MOD-1. The MOD-2 tower is a welded-steel cylindrical shell design. (See figure 2–1) This configuration leads to a "soft" flexible tower rather than "stiff" (rigid) design employed in the MOD-OA and MOD-1 design. The natural-vibration mode of the tower is at a frequency less than excitation frequency of the two-bladed rotor (that is, twice per rotor revolution). As such, the machine excites the tower vibration mode as it speeds up or slows down, but the vibration is expected to be sufficiently damped to avoid problems. The new soft-tower design appears to be very successful in reducing WT weight and therefore system costs.

Large wind turbines such as Boeing's MOD-2 have been designed for quantity production. As such, they can be expected to have substantially lower capital costs than earlier prototype models. As order rates increase, greater amounts of automated production can be introduced with attendant decreases in per-unit costs. Discounts due to quantity purchases of materials and components also contribute to cost savings. Nonrecurring engineering design, development test, and tooling costs can be spread out over a larger number of units to reduce the per-unit cost. The capital costs of prototype units (second-unit costs) are given in table 2–2.

By comparison, the one hundredth production unit of the MOD-2 is expected to cost only $686/kW (1977 dollars) installed, exclusive of land, utility interconnection, and administrative costs. The one hundredth production unit of a 200-kW HAWT of advanced design is projected to be $1014/kW [5]. Boeing reportedly predicts the cost of energy from a 25-unit cluster, installed at a 6.3 m/s (14 mph) site, to be less than 4¢/kWh (1977 dollars) as required by their contract.

Design of the MOD-2 began in August 1977 with a contract award by DOE to the Boeing Engineering and Construction Company. The objective of the program has been to develop a machine for general utility use with a cost-of-energy goal of less than 4¢/kWh (1977 dollars) when located at sites with annual wind speeds of 6.3 m/s (14 mph). At the present time there are three MOD-2s arranged in a triangular cluster "windfarm": at Goodnoe Hills, near the Columbia River near Goldendale, Washington. The site was proposed by the Bonneville Power Administration (BPA) and the Klickitat County Public Utility District of Washington. The machines are tied into the BPA network, and together they are expected to generate approximately 30 GWh of electricity annually [6]. [One gigawatt-hour (GWh) equals 10^9 watt-hours, or one million KWh.].

By way of underscoring the utility industry's interest in the MOD-2, Pacific Gas and Electric (PG&E) of northern California recently announced signing a letter of interest with Boeing to purchase such a machine for their

Wind-Energy Technology and Applications

Note: this 2.5 MW wind-turbine generator, one of a cluster of three machines that began producing power for northwest utility customers early in 1981, rises from the Goodnoe Hills near Goldendale, Washington, not far from the Columbia River. The rectangular structure atop the 200-ft.-tall tubular steel tower is called the nacelle. It is 37 ft. long and houses the drive train, turbine generator, electronics, and other equipment. The rotor is 300 ft. long. The machines were designed and built by Boeing Engineering and Construction Company under a program conducted by the U.S. Department of Energy and managed by the National Aeronautics and Space Administration. Photo courtesy of the Boeing Engineering and Construction Company.

Figure 2–1. Mod-2 Wind Turbine

Table 2–2
Cost Projections for Federally Funded Large Wind Turbines
(*second-unit costs*)

	Cost (1977 Dollars)	Rating
MOD-OA	8050 $/kW	200 kW
MOD-1	2700	2000
MOD-2	1350	2500
Alcoa	700	500

Source: J.R. Ramler and R.M. Donovan, *Wind Turbines for Electric Utilities Development Status and Economics,* June 1979, NASA TM-79170, and W.N. Sullivan, *Economic Analysis of Darrieus Vertical Axis Wind Turbine Systems for the Generation of Utility Grid Electrical Power,* Volumes 1–4, August 1979 (Sandia Laboratories Report No. SAND 78–0962).

service area. Construction at PG&E's Solano County, California, site was scheduled to start in the summer of 1981. Operation is expected to begin in early 1982. Predictions by PG&E and Boeing indicate that the annual energy production from the site (8.1 m/s) could range between 7.4 and 10 GWh annually.

Darrieus VAWT. The vertical axis configuration has the advantage of lower cost because the generating equipment is at ground level. Sullivan analyzes the economic performance of these systems [7]. Additional discussion is provided below in the context of the ALCOA program.

Private-Company Large Wind-Turbine Production Activities

Several private companies are developing and selling large WTs with rated capacities of between 100 kW and 4000 kW. Their products are of interest to utilities because they are now being interconnected and automatically controlled in utility networks. Table 2–3 lists these companies and cites the characteristics of their machines. The following paragraphs briefly summarize the status and experiences for each of the major companies.

Aluminum Company of America (ALCOA)

The Aluminum Company of America is the leading U.S. manufacturer of intermediate and large vertical-axis wind turbines (VAWTs) of the Darrieus design. They have developed a 500-kW design and are marketing it to electric utilities. (figure 2–2) The Eugene Water and Electric Board of Eugene, Oregon, as program manager for a group of electric utilities in the

Wind-Energy Technology and Applications

Note: Photograph taken at Alcoa Technical Center, Alcoa Center, Pennsylvania. Similar prototypes were installed in 1980 at Agate Beach, Oregon, and in the San Gorgonio Pass, California. Rated power is 500 kW attached at 15.7 m/s (35 mph). Machine diameter is 25 m. Photo courtesy of the Aluminum Company of America.

Figure 2–2. Prototype of Alcoa ALVAWT Model #1238229-500kW

Central Lincoln Public Utility District, and the Southern California Edison Company have purchased 500-kW machines that were installed in their respective services areas early in 1981.

In parallel with its work on the large VAWTs, ALCOA has been designing and manufacturing four low-cost, 100-kW, 17-m machines for DOE/Sandia Laboratories (see table 2–1). The first machine has been delivered to DOE's Rocky Flats test site to be used for research activities. The second machine went to the U.S. Department of Agriculture and is being used at Bushland, Texas, in connection with experimental agricultural irrigation projects. The third machine has been installed at a site on Martha's Vineyard Island off the Massachusetts coast. The last machine was planned for installation at a site on Little Equinox Mountain in southern Vermont, but recent budget restrictions within DOE have led to changes in plans. The site for the fourth machine is still uncertain. Sullivan provides an economic analysis of vertical-axis wind turbine systems [7].

Mehrkam/Energy Development Company

The Mehrkam/Energy Development Company (Hamburg, Pennsylvania) markets a six-bladed, 40-m diameter, 2-MW wind turbine designed to deliver synchronous power directly to either on-site loads or utility systems. The first installation of such a machine has benefited from the support of the local utility (Metropolitan Edison) who will be recording electric power data once prototype development problems of the WT are resolved. In addition,

Table 2–3
Key, Privately Funded, Large Wind Turbines

Manufacturer	Machine Type	Rated Power (kW)	Rated Wind Speed (m/s(mph) @ 9.1m ht.)	Rotor Diameter (m)
Alcoa Aluminum Company	V	500	15.7 (35.0)	25.0
		300	13.4 (30.0)	25.0
Mehrkam/Energy Development Company	H	225	10.9 (24.5)[a]	22.8
		45	11.8 (26.5)[a]	13.7
		2000	13.4 (30.0)[a]	39.6
Hamilton Standard	H	4000	11.2 (25.0)	77.6
Bendix/Wind Power Products	H	3000	14.3 (32.1)[a]	50.3
WTG Energy Systems, Inc.	H	200	10.6 (23.6)[a]	24.4
Boeing Engineering Company	H	2500	12.3 (27.7)	91.5

Note: V is the vertical axis, Darrieus, and H is the horizontal axis.
[a]Assume wind-shear exponent $\alpha = 0.17$ to calculate rated wind speed at 9.1-m (30 ft) height when value is provided only at hub height.

Mehrkam has sold several smaller 45- and 225- kW WTs described in table 2–3.

Hamilton Standard

The Hamilton Standard Division of United Technologies has embarked on the development of a 4-MW HAWT for the Water and Power Resources Services (WPRS) of the U.S. Department of the Interior. The machine will be installed at the agency's proposed wind-hydro site at Medicine Bow, Wyoming. Delivery was scheduled for the fall of 1981, with first rotation expected in early 1982. This is the first step in developing a 150-MW wind-turbine site whose output will be integrated with the storage and electricity-generating capabilities of the Colorado River Storage Project administered by the WPRS.

Bendix/Wind Power Products

Southern California Edison (SCE) Company recently installed a 3-MW horizontal-axis wind turbine (HAWT) manufactured by the Bendix Corporation. The machine is located at the utility's substation in the San Gorgonio pass area near Palm Springs, California. The machine is directly interconnected with the SCE network. At the present time, it has been fully assembled and is undergoing checkout and test.

WTG Energy Systems, Inc.

WTG Energy Systems, Inc. (Buffalo, New York) markets a three-bladed 24.4-m (80 ft) diameter HAWT rated at 200 kW. Currently they have completed fabrication of their third unit interconnected with an electric utility. The latest was purchased by Pacific Power and Light (PP&L).

The first machine manufactured by WTG Energy Systems is a test prototype. It is periodically interconnected with the local electric utility on Cuttyhunk Island off the coast of Massachusetts in connection with machine-test activities. During tests it has carried a major portion of the island's electrical load.

Boeing Engineering and Construction Company

Boeing developed the MOD-2 under DOE contract and now markets this machine for electric-utility applications. As discussed earlier, and shown on table 2–1, the MOD-2 is considered an important development. We selected the Boeing MOD-2 for the simulation study (chapter 5) because:

1. A three-unit windfarm will be on line at Goodnoe Hills, Washington, in 1981.
2. This is the first utility size WECS to be erected in a windfarm configuration.
3. This is one of the primary utility-scale wind turbines that is expected to see application by electric utilities in the early 1980s.

Our selection for modeling does not in any way imply endorsement, nor does it reflect any evaluation of its performance and cost in comparison with other commercially available large wind turbines.

Overview of Major-Utility Wind Programs

Generally, the electric-utility industry has been slow in taking the initiative in fostering the use of renewable-energy sources such as the wind to generate electricity. However, some early innovators such as Hawaiian Electric Company (HECO), Southern California Edison (SCE), and Pacific Gas and Electric (PG&E) have shown a serious interest in integrating this resource into their mix of generating equipment. (It is noted that HECO contracted with third parties to purchase power rather than invest on their own.)

Hawaiian Electric Company (HECO)

Hawaiian Electric Company and Windfarms Limited (WFL), a San Francisco-based windfarm developer, signed contracts in mid-1980 making it possible to proceed with the world's largest wind-electric project that will be located at Kahuku Point, Oahu, Hawaii. This is also the first large wind-energy project undertaken jointly by an investor-owned utility and private-sector financing. Current plans call for the installation of 5 MW by early 1983. Once begun, 4-MW increments will be added every month or so until a full 80 MW of windfarm to save the utility about 800,000 barrels of oil annually. This figure is derived from an early HECO analysis that estimated that 32 Boeing MOD-2 (2.5 MW) wind turbines could generate 490 million kWh annually thereby providing Oahu with 8.5 percent of its 940-MW peak load.

Besides Hamilton Standard, two other companies are participating in the financing of the project. Bechtel International, responsible for the project's engineering, and Karlskronavarvet, Hamilton Standard's partner in a prior Swedish WT project, are assuming an equity position in addition to providing guarantees on the performance of the project. In total, the project has been reported to cost $320 million with $100 million being contributed by the three vendors.[2]

The terms of the purchase agreement establish a minimum of 6¢/kWh with no upper limit on the price HECO will pay for WFL's electricity. This price is keyed to the average price of imported oil and amounts to about 90 percent of HECO's "avoided cost." (See PURPA chap. 4, p. 80) In addition to purchasing all the electricity generated by the 80-MW array, Hawaiian Electric will maintain and operate the machines for WFL once the project goes into commercial operation. HECO may also be interested in eventually owning wind-energy facilities. One of the terms of the purchase-power agreement gives HECO the right of first refusal should WFL eventually decide to sell the facility.

Southern California Edison

Southern California Edison plans to add 130 Mw of wind energy to its grid by 1990. This number represents 6.3 percent of the 1900 MW that SCE feels can be provided by renewable-energy sources. Their plans call for the remainder to be captured for other resources: geothermal, 420 MW; cogeneration, 300 MW; hydroelectric, 620 MW; fuel cells, 130 MW; and solar-thermal and photovoltaics, 310 MW [8].

Although innovative by industry standard, SCE's plans are conservative when compared to the potential for wind power available in its service area. Zambrano estimated that the San Gorgonio Pass Area has an annual wind-power density of 600 to 1200 W/m [9]. These values are based on average wind speeds of 22 to 28 mph. The same study also estimates that an array of 50, 4-MW wind turbines could produce 516 million kWh annually in a 32-mph wind.

As a first step in attempting to meet the goals set for itself, SCE has purchased a large, three-bladed, 3-MW Bendix wind turbine and an intermediate size 500-kW Darrieus vertical-axis machine manufactured by ALCOA. In so doing, SCE has become the first investor-owned utility in the country to purchase and operate a megawatt-size machine. (The 500-kW ALCOA WT was damaged in a wind storm during commissioning activities early in 1981 and is not now in service.) The Bendix turbine is designed to reach rated power in a 40-mph windspeed in order to maximize power output for the San Gorgonio Pass wind regime. SCE estimates that the Bendix machine can generate 6 million kWh annually, thereby displacing approximately 10,000 barrels of fuel oil yearly.

Pacific Gas and Electric Company

As mentioned earlier, PG&E expects to have a Boeing MOD-2 wind turbine operating by 1982. The significance of this project is that it is the essential first step for PG&E in developing the large wind resource in their service areas, which includes virtually all of northern California.

By way of follow-up to this step, PG&E has recently announced signing a letter of intent with Windfarms Limited to undertake the joint development of a 320-MW windfarm to be located in Solano County near Fairfield, California. WFL will arrange for purchase of the wind turbines (probably a mixture of MOD-2s and smaller units). PG&E will be supplying the site, the Swett Ranch, which has been purchased expressly for the Windfarms project. The timetable for this project is uncertain at the present time, but it is likely to take several years to complete.

Third-Party Wind-Turbine Enterprise

To promote the use of renewable-energy sources, Congress passed the Public Utility Regulatory Policies Act (PURPA) in 1978 (see chapter 4 for a detailed discussion of PURPA). Among its provisions PURPA requires that utilities purchase power at "just" rates from qualifying "small power producers" (the "third parties") who employ renewable-energy resources to generate power. In addition to purchase, the utilities are obligated to transmit power on behalf of the third party and to establish a fair rate for doing so.

PURPA exempts certain types of small power-production facilities such as wind clusters (that is, windfarms) from a number of provisions of the Federal Power Act, the Public Utility Holding Company Act, and from state laws governing the setting of utility rates. The intent of PURPA has been to create incentive conditions whereby independent or third-party producers of electricity can operate without reference to traditional utility regulations. The potential for financial success of such an enterprise has begun to attract the attention of investors who are willing to assume the financial and technical risks necessary to make windfarming a reality.

To date, two windfarm developers have publicly announced projects, although several others are currently planning or negotiating projects. One is U.S Windpower (USW) of Burlington, Massachusetts, and the other is Windfarms Limited (WFL) of San Francisco. In each instance, purchase-power agreements have been negotiated. Financing and engineering efforts are currently underway for both developers.

Each developer has chosen a different approach to raising capital for development work. USW's approach is to attract individual investors interested in sheltering other taxable income. The investor can take advantage of provisions of recent incentive legislation, 1980, which sets

Wind-Energy Technology and Applications

out tax credits and depreciation allowances available to purchasers of renewable-energy equipment. (See chapter 4 for discussion of the legislation.) In this approach the individual investor purchases a wind turbine from USW, and USW arranges for it to be installed in one of the company's windfarms. The investor next enters into a management contract with USW to service the equipment and to arrange sale of the power to the utility. USW receives income from the sale of the wind turbine and the performance of the service contract. The investor's income is derived from the income produced by the sale of the power and from the benefits of the tax shelter.

Windfarms Limited, on the other hand, arranges for debt and equity financing from such conventional sources as insurance companies and investment banks. They subcontract for the design of the overall windfarm system, selection of wind turbines, supervision of overall construction, and installation and operation of the equipment. They negotiate for a power-purchase agreement, maintenance contract, and so on.

Benson, among others, argues that the incentives for third-party energy generation under PURPA are a disincentive to the electric-utility industry itself [10]. While the utility is free to invest in wind energy, it cannot take advantage of the tax incentives available to a third party. This is viewed by Benson as effectively denying the utility the incentive needed to take the risk on a new technology, leaving equipment selection, operation, and maintenance of the machinery to third parties that may lack the technical experience of the utility.

However, it does appear possible that, through negotiated agreements such as that between WFL and HECO, the utility can be retained for key roles that they can best fulfill, while allowing the windfarm developer to take advantage of the incentives in the short term.

Foreign Wind-Energy Programs

The United States is not alone in the development of large wind machines for utility application. Several other countries have made a significant investment into wind-energy development, as will be briefly described here.

Denmark

The Danish wind-energy program was initiated in 1977 and has continued for the past four years. Working with a total budget of approximately $8M over a three-year period, they have the following objectives:

> Assess the feasibility of interconnecting large wind turbines with a utility grid

Develop improved estimates of large wind-turbine construction costs

Gain insight into expected reliability estimates, maintenance costs, and estimated lifetimes

In an effort to achieve these goals, the Research Association of the Danish Electricity Supply Undertakings (DEFU), which manages the project, has concentrated on building and/or testing large wind turbines primarily located at Gedser and Nibe, Denmark. The Gedser WT was refurbished in 1977 (after ten years of inactivity) in connection with a technical, acoustic-noise, and television-interference (TVI) test program partly funded by the U.S. DOE. The program was completed in 1979 and the results have been reported by Lundsager et al. [11]. Many of the results were similar to those found in the U.S. DOE test programs with the MOD-O/OA and MOD-1 WTs. In the main, the Danish wind-energy program has focused on the design, construction, and testing of two new 630-kW three-blade HAWTs at Nibe, Denmark. The overall machine design and manufacture was carried out by a number of firms. The VØLUND Company of Viborg, Denmark, provided the blades while Danalith Ltd. constructed the building.

The two machines that are nearly identical became fully operational in late 1980 and early 1981, and hence only a limited amount of experimental data are available at present. Pederson and Nielsen report that initial results show fundamental compliance with anticipated power output and loading on the structural components [12].

Sweden

Swedish meteorological conditions are dominated by the global wind belt known collectively as the "prevailing westerlies." Hunt reports that especially windy areas are known to exist along the West Coast of Sweden, in many areas in the southern province of Skane, on the islands of Oland and Gotland, and in northern Uppland [13]. A recent article reports the results of a study by the Swedish Secretariat for Future Studies [14]. The study indicates that if a large number of available sites with wind speeds exceeding 6 m/s (13.4 mph) were fully utilized, 14.8 GW of wind-turbine capacity would yield 30,000 GWh annually. In an effort to realize this potential, the Swedish government has undertaken a program of wind-turbine development similar to that in the United States.

In order to gain design and construction experience as early as possible, a prototype wind turbine was ordered from Saab-Scania Company in early 1976. The two-bladed wind turbine measures 18 m (59 ft) in diameter. It is located at Kalkugnen near Alukarleby on the Baltic coast. Testing began in 1977 and has thus far proceeded according to plans.

Wind-Energy Technology and Applications

In September 1979 the Swedish National Board for Energy Source Development selected a proposal for a megawatt-sized wind turbine submitted jointly by Karlskronavarvet AB of Sweden and Hamilton Standard of the United States. Hamilton Standard has designed the rotor, manufactured two 125-ft (38.1 m) fiberglass blades, and conducted a computer analysis of the entire system. Karlskronavarvet will design the turbine's 260-ft (79.3 m) tower, its foundation, and the assembly that houses the generator. Plans call for the Swedish company to erect the machine at Maglarp on Sweden's south coast near the city of Malmo in the fall of 1981. The wind turbine features a downwind free-yawing generator on a tubular steel tower. With a hub height wind speed of 13.8 m/s (31 mph), it will generate 3 MW of power. The turbine is designed to operate in the wind-speed range of 4.5 m/s (10 mph) to 22.4 m/s (50 mph). Hamilton Standard estimates that the wind turbine will produce approximately 8 million kWh annually at a site with an annual average wind speed of 6.3 m/s (14 mph).

The Swedish government has also awarded a second contract for a 2-MW wind turbine to Karlstads Mekaniska Werkstat AB (KMW) jointly with ERNO Raumfahrtechnik of West Germany. KMW, who is responsible for coordinating the entire project, developed the hub, machinery, nacelle, and control system, and that company will do the erection. ERNO, a subsidiary of the German VFW-Fokker group has been responsible for the design, analysis, and construction of the rotor assembly. The wind turbine is a two-bladed, upwind machine with a 75-m (248 ft) diameter rotor. A listing of the major specifications for this wind turbine as well as the Hamilton Standard machine are given in table 2–4.

West Germany

In the summer of 1976, the Grosse Windenergie-Analage (Growian) program was begun as West Germany's principal effort in developing utility-scale wind turbines. The wind turbine that has evolved from this program has been designed with specific reference to wind sites along the North German coast where wind speeds typically average 6 m/s (13.4 mph) measured at 10 m above the ground. With such a resource, their 3-MW wind turbine has been estimated to yield 12 GWh annually (Korber and Thiele) [15]. The machine that will supply this power is a two-bladed, 100-m (329-ft) diameter downwind design that reaches rated power in a 12-m/s (26.5 mph) wind speed. Power production begins at a wind speed of 6.3 m/s (14 mph), and safety shutdown occurs when the wind reaches 24 m/s (54 mph). The machine is scheduled for commissioning sometime in mid to late 1982. Currently prototype blades have been fabricated and are undergoing structural testing.

Table 2-4
Swedish Wind Turbines

Characteristic	KKRV/HS	KMW/ERNO
Number of blades	2	2
Blade diameter	80 m (261 ft)	75 m (254 ft)
Speed, rpm	25	25
Rotor, relative to tower	Downwind	Upwind
Hub type	Teetered	Rigid
Hub height	79 m (260 ft)	77 m (260 ft)
Tower type	Welded steel shell	Reinforced concrete
Tower diameter	3.8 m (12 ft)	4.5–9.5 m, tapered
Generator type	Synchronous AC	Induction
Grid connection	55 kV	30 kV
Orientation drive	Free yaw	Yawed via hydraulic motors
Rated power	3 MW	2–2.35 MW
Wind speed at centerline		
Cut-in	5.8 m/s (13.0 mph)	6.03 m/s (13.5 mph)
Rated	13.6 m/s (30.4 mph)	13.4 m/s (30.1 mph)
Cut-out	22.7 m/s (50.8 mph)	21.0 m/s (47 mph)

Source: Private communication with the National Swedish Board for Energy Source Development.

United Kingdom

The British Isles, particularly Scotland and Wales, possess an abundant wind resource. The Electrical Research Association has maintained an interest in this resource since the 1950s having, over the years, identified nearly 1500 potential wind-turbine sites on hilltops in Great Britain [16]. One such location is Burgar Hill on the Island of Orkney, located off the northern coast of Scotland where the U.K. Department of Energy will erect a 3-MW wind turbine by late 1983 or early 1984. With this project, the United Kingdom becomes the third European country, after West Germany and Sweden, to develop a megawatt-size machine.

A group of three companies, headed by Taylor-Woodrow Construction, Ltd. will build the machine with funds provided jointly by the U.K. Department of Energy and the North of Scotland Hydro Electric Board. Taylor-Woodrow will do the conceptual design, systems engineering, and dynamic analysis. It will also construct the foundation and the concrete tower. GEC Power Engineering Ltd. will design, manufacture, assemble, and test the generator, transmission, and nacelle assembly. British Aerospace will design, build, and test the rotor assembly.

The wind turbine will have a 60-m (197 ft) diameter blade mounted atop a 46 m (150 ft) high concrete tower. Startup is at 7 m/s (15.7 mph), reaching the rated 3 MW, rated at 17 m/s (38 mph), with cut-out at 27 m/s (60 mph). With 12 m/s (27 mph) at hub height, Taylor-Woodrow estimates that the wind turbine could generate 11 million kWh annually [17].

Canada

Wind power in Canada has great potential in both the grid- and non-grid-connected parts of the country. In isolated regions such as the Northwest Territories where diesel generators are the only power sources, wind turbines are already cost effective. The same can also be said for the southern prairie provinces, and the eastern coastal regions around the Gulf of Saint Lawrence including the maritime provinces. For example, a study of the wind resources of Prince Edward Island indicates that 10 MW of installed wind-power capacity would be economic in the first year of service [18].

One of the windiest locations in Canada is on the Magdalen Islands located in the middle of the Gulf of Saint Lawrence. It was here in 1977 that the National Research Council (NRC) and Hydro-Quebec (the local utility) erected a 230-kW Darrieus WT manufactured by DAF-Indal Corporation. It operated, interconnected with the grid, for only about 200 h before it was damaged by a windstorm in the summer of 1978. Following the accident, the machine underwent redesign and refabrication and was returned to service in March 1980. Since then it has accumulated approximately 100 h of intermittent operation in support of research activities.

In other activities, NRC and Hydro-Quebec announced in early 1981, plans for the development of a 64-m (210 ft) diameter Darrieus wind-turbine called "Project Aeolus." The machine is tentatively rated at 3.8 MW in a 14.3-m/s (32 mph) wind regime. When completed in 1983, the Aeolus will be the world's largest vertical-axis wind turbine standing 110 m (360 ft) high. The actual site for the machine has not been determined, but several locations along both shores of the Saint Lawrence River and at the Magdalen Islands have been suggested. At a site with an 8.1-m/s (18-mph) wind speed, the Aeolus is estimated to generate 6.1 million kWh annually.[3]

Off-Shore Wind-Energy Projects

The generation of electricity by using wind turbines positioned off-shore has some advantages. It should be possible to optimize WT designs for higher cut-out speeds, thereby making it possible to take advantage of the generally higher wind speeds occurring off-shore. Also questions of environmental impact, land-use restrictions, and television-interference problems are largely avoided.

However, these advantages can be realized only at greater costs in terms of installation, maintenance and transmission of the power produced. A Westinghouse study estimated that a 500-MW off-shore array could produce electricity at a busbar cost of between 6.4 and 17.6¢/kWh [19]. This compares with 3 to 5¢/kWh for electricity produced from a 25-unit land-

Table 2-5
Estimated Displacement of Oil Using Large WECS

No.	Electric Utilities Using More than 5 Million Barrels of Oil in 1978	Oil Consumed (10^3 bbl) 1978	Percent Oil Consumed in ≥ 300 W/m^2 Wind Regime (10^3 bbl) 1978	WECS[a] On-Line Capacity (MW) 1980	WECS Oil Displacement (10^6 bbl) 1980
1	Southern California Edison Co.	45,203	100	1,260	7.29
2	Consolidated Edison Co. of N.Y.	39,255	100	705	4.08
3	Middle South Utilities, Inc.	39,102	30	350	2.02
4	Florida Power & Light Co.	36,452			
5	Pacific Gas & Electric Co.	28,821	100	1,362	7.88
6	Virginia Electric & Power Co.	26,485	100	820	4.74
7	New England Electric System	21,261	100	325	1.88
8	Long Island Lighting Co.	21,027	100	315	1.82
9	Public Service Electric & Gas Co.	20,538	100	695	4.02
10	Florida Power Corp.	16,331	0	—	—
11	Niagara Mohawk Power Corp.	14,214	100	576	3.33
12	Boston Edison Co.	14,067	100	213	1.23
13	Philadelphia Electric Co.	13,276	100	595	3.44
14	Los Angeles Department of Water and Power	13,207	100	435	2.52
15	Northeast Utilities	13,159	100	406	2.35
16	Gulf States Utilities Co.	12,688	0	—	—

17	Commonwealth Edison Co.	12,665	0	—
18	Potomac Electric Power Co.	11,555	100	2.26
19	Consumers Power Co.	11,073	100	2.79
20	Pennsylvania Power & Light Co.	11,052	100	2.86
21	San Diego Gas & Electric Co.	10,813	0	—
22	Jacksonville Electric Authority	10,729	0	—
23	Baltimore Gas & Electric Co.	10,653	100	2.16
24	Hawaiian Electric Co., Inc.	10,168	100	0.65
25	The Detroit Edison Co.	8,958	0	—
26	United Illuminating Co.	8,664	100	0.58
27	Delmarva Power & Light Co.	8,439	100	0.87
28	New England Gas & Electric Assn.	7,513	100	0.37
29	The Southern Company	6,544	0	—
30	Tennessee Valley Authority	6,230	0	—
31	General Public Utilities Corp.	5,894	100	3.59
32	South Carolina Electric & Gas Co.	5,535	0	—
33	Central Hudson Gas & Electric Corp.	5,329	100	0.38
	Subtotals (82.9 percent)	526,900	71	63.11
	Remaining (17.1 percent)	108,918	—	—
	U.S. totals	635,818	59	63.11

Source: Johanson, F.E, Goldenblatt, M., Marshall, R., and Tennis, M. *Markets for Wind Energy Systems*, AIAA/SERI Wind Energy Conference, April 9–11, 1980. Reprinted with permission.
[a] Assumed to be 10 percent of 1980 peak load in ≥ 300 W^2/M^2 wind regime.

based MOD-2 array. The important cost-escalating factors were higher wind speeds, water depth, and more stringent survival-design conditions. The adverse impacts of corrosion and fouling also add to the costs. Distance to shore had little added effect when compared with other cost factors. The same study also estimated that the electrolysis of water into hydrogen and its transmission to shore and reconversion to electrical energy is six times more expensive than submarine cables. Overlaying many of these technical constraints are concerns over potential legal and institutional risks and complications due to liability in case a floating unit breaks free and drifts into shipping lanes or if a ship should collide with off-shore WTs. As a result, the off-shore applications of wind turbines are not expected to occur nearly as soon as those at selected sites on-shore.

Ultimate Utility Wind-Energy Potential

Johanson et al., using rough first-order estimating methods, without regard to the availability of land and power lines, computed the amount of oil that would be saved by windpower if all utilities with access to a minimal level of wind resource were to install wind at the 10-percent penetration level, based on 1980 peak load [20]. This would entail almost 12,000 MW of on-line capacity.

They demonstrate that the 10-percent penetration level can be expected to displace 63 million barrels annually, or about 10 percent of all oil consumed by electric utilities in 1980. Their detailed estimate is given in table 2–5.

We note that a 10-percent penetration should not necessarily be considered an upper limit unless operational problems limit further penetration. In addition, as will be seen in chapter 6, wind energy can profitably displace coal as well as oil fuels. Finally, where hydroelectric power with storage is available, wind power can effectively increase the firm yield of the hydropower system.

Notes

1. Also see SERI's *Semi-Annual Reports of the Photovoltaic Research Branch* (SERI/PR-611-737) and DOE's *National Photovoltaic Program* (DOE/ET-0105-D).
2. Arthur D. Little, Inc., personal communication.
3. Arthur D. Little, Inc., private communication.

References

1. Braun, G.W. *The U.S. Department of Energy Solar Thermal Systems Program—An Overview Presentation*, August 1979, SERI/SP-733-526.
2. Stobaugh, R., and Yergin, D. *Energy Future.* New York: Random House, 1979.
3. Sandia Laboratories. "Photovoltaic System Definitions and Development." Project Integration Meeting. Albuquerque, N. Mex., October 21–23, 1980. Sandia Laboratories Report No. SAND 80-2374.
4. Vachon, W.A. *Large Wind Turbine Generator Performance Assessment, Technology Status Report No. 1*, January 1980, EPRI Report AP-1317.
5. Ramler, J.R., and Donovan, R.M. *Wind Turbines for Electric Utilities Development Status and Economics,* June 1979, NASA TM-79170.
6. "Going With the Wind." *EPRI Journal* 5, no. 2 (March 1980).
7. Sullivan, W.N. *Economic Analysis of Darrieus Vertical Axis Wind Turbine Systems for the Generation of Utility Grid Electrical Power*, Volumes 1–4, August 1979. Sandia Laboratories Report No. SAND 78-0962.
8. "SCE To Air 360 MW Wind Power by 1990." *Wind Energy Report*, October 1980.
9. Zambrano, T.G., Walker, S.N., and Baker, R.W. *Wind Energy Assessment of the Palm Springs—Whitewater Region: Executive Summary.* Prepared for the California Energy Commission by Aero Vironment Inc., February 1980.
10. Benson, C.C. *Legislative, Regulatory and Institutional Barriers to Electric Utility Participation in Energy Conservation and Renewable Resource Programs* (a position paper). Dallas Power & Light Company, Edison Electric Institute, August 13, 1980.
11. Lundsager, P., Frandsen, S., and Christensen, C.J. *Analysis of Data from the Gedser Wind Turbine 1977–1979.* Risø National Laboratory, Denmark, August 1980.
12. Pederson, B.M., and Nielson, P. "Description of the Two Danish 630 kW Wind Turbines, Nibe-A and Nibe-B, and Some Preliminary Test Results." *Third International Symposium on Wind Energy Systems.* Copenhagen, Denmark, August 1980, pp. 223–238.
13. Hunt, V.D. *Windpower, A Handbook on Wind Energy Conversion Systems.* New York: Van Nostrand Reinhold, 1981.
14. "Hamilton Standard Gels Swedish WECS Contract." *Wind Energy Report*, September 1979.

15. Korber, F., and Thiele, H.A. "Large Wind Energy Converter—Growian 3 MW." *Large Wind Turbine Design Characteristics and R&D Requirements.* Workshop held at Lewis Research Center, Cleveland, Ohio, April 24–26, 1979.
16. Golding, E.W. *The Generation of Electricity by Wind Power.* New York: Wiley, 1976.
17. "Great Britain to Build 3 MW, 250 kW WECS." *Wind Energy Report*, January 1981.
18. Templin, R.J. "Wind Energy." *Proceedings of the Third Canadian National Energy Forum, Halifax,* April 4–5, 1977.
19. Kilar, L.A. *Design Study and Economic Assessment of Multi-Unit Off-Shore Wind Energy Conversion Systems Application.* Final report, vol. 1: Executive Summary. Prepared for the U.S. Department of Energy by Westinghouse Electric Corporation, Pittsburgh, Pa., June 1979.
20. Johanson, E.E., Goldenblatt, M., Marshall, R., and Tennis, M. *Markets For Wind Energy Systems.* AIAA/SERI Wind Energy Conference, April 9–11, 1980.

3 Methodology of Study

We begin with the assumptions that large-scale wind power for utilities will be economically feasible in regions with an appropriate wind resource and that wind-generating machines will be commercially available. We then explore the conditions under which a utility will make a major investment in wind energy (say, 1000 MW) and identify perceived barriers to commitment. Finally, we identify federal and state policies that appear to be practical, in order to test the hypothesis, do these policies motivate the electric-utility industry to invest in wind power?

A major feature of the methodology is the use of a "synthetic utility" as a proxy for many utilities with analogous characteristics. The hour-by-hour operation of this utility, over a fifteen-year planning horizon, is simulated using a production-cost model: without a wind project and then with a postulated wind project. The stream of annual operating costs in each case becomes input to a corporate financial model of the synthetic utility. Thus the impact of the wind project on the utility's financial condition can be measured, and the factors that enter into an investment decision can be articulated.

Finally, by altering the financial model to correspond with each federal- and state-policy alternative, we measure the impact of each policy on the financial performance of the utility in the presence of the wind investment. Each hypothesis is thereby tested, using specific financial performance criteria. We then evaluate which policies are likely to be effective. The methodology uses two major computer tools: a production-cost model and a corporate-finance model. It also employs data tapes containing wind data, load data, other utility data, and a program to predict wind energy from a wind generator.

The methodology encompasses the capability to evaluate the engineering-economics of the wind project as well as the impact of the project on the company's financial performance. While wind is used as the case study, the methodology is general in nature and can be used to evaluate any investment into solar-electric energy or into energy conservation.

We assumed that each candidate investment is credited with fuel savings only and that the expansion plan of the utility is not altered. This is realistic for wind energy in today's decision-making climate in which utilities generally do not recognize an intermittent energy source as contributing to firm capacity. There are methods for estimating a capacity credit ranging from the simple (for example, computation of improved system reliability) to the extremely complex (for example, use of optimality algorithms to predict the utility's future equipment mix for each scenario of renewable energy or conservation). Capacity-credit algorithms that can be used as practical tools for rapid evaluation of wind and other technologies can be developed, but it is beyond the scope of this study to do so.

We have also assumed the absence of electric storage based on today's perception of excessive system costs. The methodology can be extended to include simulation with storage and the estimation of the appropriate capacity credit, but this is similarly beyond the scope of this study.

This chapter is presented as follows:

Background and Literature Review. The basic project-evaluation tools that have been applied to wind energy are explained, and a review of literature on the subject is provided.

Simulation. The specific models we use herein are described: (1) to simulate the reduction in operating costs induced by wind energy and (2) to simulate the year-to-year balance sheet and other financial reports of the utility.

Background and Literature Review

A solar-electric installation may pay for itself in two ways:

By reducing fuel and related operating costs to the utility (fuel savings)

By justifying the elimination or delay of capital expenditures that would otherwise be required (capacity credit)

Fuel-Savings Methodology

The fuel-savings justification is widely accepted in the utility industry as a basis for evaluating investments into solar-electric technology. The capacity credit is considered somewhat controversial, as utilities, in general, are not willing to adjust their conventional expansion plan around a stochastic (that is, intermittent) source technology. Nevertheless, the capacity-credit concept has a sound theoretical basis and may sometimes equal or exceed fuel savings in economic importance. This will be illustrated later in this chapter.

Fuel savings are obtained when the utility reduces the operating level of an oil, coal, or nuclear plant in response to a reduced load induced by the

Methodology of Study

presence of wind energy on the grid. The value of the fuel saved depends on the particular generator that is backed down to a lower operating level and the unit value of the fuel at that time. Thus the value of the fuel displaced by the presence of wind depends on:

The hour of the day

The day of the week

The season of the year

Also, since the equipment mix of a utility changes from year to year, the fuel savings depends on the year itself. The most accurate methodology for estimating the fuel savings, given these variations, is a simulation of the hour-by-hour equipment dispatch by the utility to meet the load, taking account of scheduled and forced outage events of all equipment controlled by the utility. By simulating the fuel used under a base-case simulation without wind and then resimulating in the presence of wind, the cumulative reduction in fuel costs can be obtained for any time period of interest. This kind of simulation is usually accomplished using a utility *production-cost model.*

Figure 3–1 shows the *production-cost model* at the heart of a typical methodology for solar-electric value analysis. A *wind-power model* provides simulated or historical values of wind speed each hour. A *utility-load model* specifies the electric energy that the utility must generate each hour of the year. A *utility equipment-mix model* specifies for the current year the list of generating equipment that can be dispatched at a given hour. The production-cost model integrates all this information and selects the minimum-cost combination of available equipment to satisfy the load each hour. The difference in production cost, with and without wind, is the current year's savings. If the utility's load and/or equipment mix are expected to change over the years of interest, each subsequent year needs to be simulated. The series of annual savings, suitably escalated by an estimate of future fuel-price increases, becomes the input to a *life-cycle model,* which is used to compute the revenue requirements, the breakeven cost, or the levelized cost.

The use of production-cost models to estimate fuel savings induced by solar electric are reported on by JBF [9, 12], Lindley and Melton [7], and others [8, 11]. Some studies use an assumed-capacity factor of the wind machine, and develop a simplified estimate of fuel displaced using a utility load-duration curve [21, 22]. The General Electric Company (GE) maintains a very detailed production-cost model, which is commercially available to utility customers [18]. The GE model is described later in this chapter.

Percival and Harper published computer programs for general use in simulating the value of fuel savings and determining the breakeven value

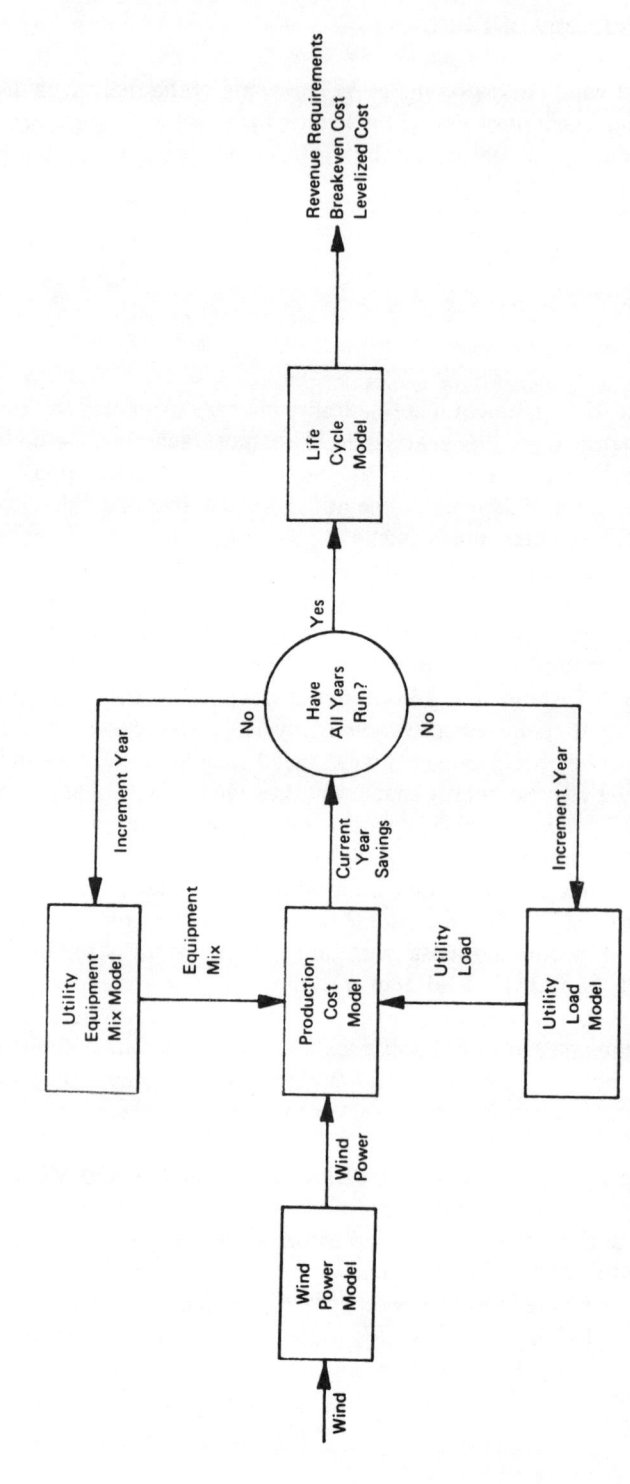

Figure 3-1. Typical Methodology for Solar Electric-Value Analysis (Simplified to show fuel saving only—capacity credit not shown)

Methodology of Study

[26]. Their work stresses enhanced capability to simulate wind energy and sensitivity testing with breakeven cost as dependent variable, but it requires the user to supply his own production-cost and utility-expansion models.

Life-Cycle Model

Most studies of the value of wind or other solar-electric power have used a life-cycle model to obtain the breakeven cost. Some calculate the levelized cost. These are defined as follows. "Breakeven cost" is the present value of the stream of savings induced by an investment, using the utility's discount rate. "Levelized cost" is the uniform stream of annual payments that has the same present value as the stream of required revenues, using the utility's discount rate (expressed in $/kWh).

Thus breakeven cost measures the amount the utility should be willing to invest while earning exactly the discount rate on the investment. Levelized cost expresses the amount that the utility has to charge its customers to recover a given investment while earning exactly the discount rate. While breakeven or levelized costs are equivalent for purposes of comparing the value of projects, the breakeven cost has been preferred because one does not need to know the capital cost of the facility. One can readily compare breakeven cost to different estimates of future installed cost. Levelized cost, on the other hand, requires that the installed cost be known. The basic equations are given here.
For breakeven cost:

PV = present value of N years of savings

FCR = utility's fixed-charge rate on capital

PWF = utility's present-worth factor

BE = breakeven cost

$$BE = \frac{PV}{(FCR \cdot PWF)} \quad (3.1)$$

CRF = capital-recovery factor

τ = income-tax rate

β = other taxes and insurance as coefficient of capital cost

N = book life of project

$$FCR = \frac{1}{1-\tau}\left[CRF - \frac{\tau}{N}\right] + \beta \quad (3.2)$$

For levelized cost:

CI_{pv} = present value of capital investment

OM_{pv} = present value of operating and maintenance

LC = levelized cost

MWH = megawatt hours generated per year

$$LC = \frac{FCR \cdot CI_{pv} + CRF\,(OM_{pv} + FL_{pv})}{MWH} \qquad (3.3)$$

To account for a delay between the time money is invested and the time power is actually produced:

g = inflation rate

d = delay time (years)

α = adjustment factor

$$\alpha = [1 + g]^d \qquad (3.4)$$

Dividing the right side of equation 3.1 by α decreases the breakeven cost. Multiplying equation 3.3 by α increases the levelized cost.[1,2]

The California Institute of Technology provides the detailed theoretical basis for computing FCR and LC [1]. Some of the earlier studies of wind-energy applications used the levelized cost per kilowatt hour of output [2, 3, 4], and Johanson reports on cost of wind energy based on studies published in 1975–1977 using levelized cost in mills per kilowatt hour, as an index [5]. The range was 2.3–3.5¢/kWh. Since the breakeven value and the installed cost can be expressed in terms of levelized cost per kilowatt hour, use of the present-value or the annualized-value convention will yield the same decisions. The present-value method is generally preferred since the levelized-cost concept can be misunderstood, for example, when people seek to compare the levelized cost with utility rates or with production costs.

Johanson later reports on the results of several independent studies [7, 10] of wind energy for utilities, each of which uses the present-value form of breakeven cost [6]. His results are shown on table 3–1 for six utilities of diverse equipment mix. These are compared with our results, based on simulation presented in chapter 5. This simply shows that our simulation is consistent with results reported in studies of the value of wind energy. Additional studies, using the present-value breakeven cost, are noted [12, 13, 14].

All of the just-cited studies deal essentially with situations in which wind energy displaces fossil fuel or, in the case of the California aquaduct,

Methodology of Study

Table 3-1
Summary of WECS Value Analysis (based on fuel savings)

Utility	Cost of Capital (Percent)	Planned Equipment Mix (Percent) Oil/Coal/Nuclear/Other	Fuel-Escalation Rate (Percent)	WECS Capacity Factor	Average Wind Speed (30 ft)(mph)	Year 2000 Breakeven Cost ($Kw) (1977 Dollars) Original	Normalized
New England Power Pool	11	34/2/55/9	9	.30	11.2	800	1110
Niagra Mohawk	10	30/31.5/13.5/25	6	.36	10.1	490	1200
Kansas Gas & Electric	10	28/57/15/0	6	.45	12.1	360	760
Northwest Power Pool (Coast Site)	10	2/13/31/53	6	.51	17.9	420	973
Northwest Power Pool (Gorge Site)	10	2/13/31/53	6	.45	16.3	335	760
Hawaiian Electric	11.7	100/0/0/0	9	.59	18.0	1900	2030
Southern California Edison	9	53/17/25/5	6.7	.26	—	380	1360[d]
ADL Synthetic Utility[a] Class 4	10	40/36/24/0	8	.27	12.8	772	1110
ADL Synthetic Utility Class 6	10	40/30/24/0	8	.34	15.0	880	1090

Source: Edward E. Johanson, JBF Scientific Corporation, *Synthesis of WECS/Utility Integration Studies: Centralized and Dispersed, Proceedings of the Fourth Bienneal Conference on Wind Energy Conversion Systems*, October 29-31, 1979.
[a]Simulation by Arthur D. Little, Inc., as in chapter 6 herein.
[b]Lifetime equals thirty years except Southern California Edison, which equals twenty years.
[c]Normalized for 0.40 capacity credit and to 11 percent cost of capital; no capacity credit unless otherwise noted.
[d]Includes capacity credit.

electrical energy on the grid. A major application of wind, however, is associated with the management of hydroelectric-power systems in which there is an opportunity to store the additional energy provided by the wind on a managed basis [15, 16, 17]. Benefits accrue from increasing the firm and unfirm electric-energy yield of the hydroelectric system. Additional benefits are derived from increasing the average storage, which provides additional head on the turbines.

Capacity Credit

The introduction of any new equipment into a utility, including wind machines, increases the utility's reliability. This is measured by the classic "loss-of-load-probability" (LOLP) calculation. When planning for utility expansion against load growth, a planner seeks to select a sufficient reserve margin (electric-generating capability above the peak load), such that the LOLP does not exceed a stated value. Thus, it is argued, the introduction of wind energy increases the reliability (decreases the LOLP) and allows the utility to spend less on new equipment in the future.

To rigorously account for capacity credit is methodologically complex. Referring to figure 3-1, it means going through the generation-planning procedure to find the optimal utility-equipment mix for each proposed wind-energy configuration. An approximation can be made without replanning the utility's equipment mix by simply computing the LOLP each year and estimating the value of new equipment, which can be removed to restore the target LOLP.

Johanson and Goldenblatt [13] and JBF [12] develop both approaches to capacity credit, calling them "reoptimized-mix mode" and "fuel-saver mode," respectively. In these papers capacity credit is very small at low penetration of wind in the New England Gas and Electric Association (NEGEA) grid in southeastern Massachusetts. At very high penetration, capacity credit accounts for as much as 15-20 percent of the breakeven value. Johanson further explores this issue across several utilities, for which the fuel displacement and capacity credit of the initial WECS (year 2000 installation) is determined, as shown on table 3-2 (approximated from published graph) [6].

Thus, theoretically the capacity credit is very important and can represent a major portion of the economic justification of a wind-energy project. Yet Johanson also reports, "A number of utilities have indicated that they would not be able to accept the capacity credit concept until the WECS reliability and performance had been verified. Furthermore, the capacity credit would not be observed today, in some utilities, because of the current cost of peaking fuels" [6]. While not explored in the paper, this would follow

Table 3-2
Capacity Credit Estimates

Utility	Total Value of WECS ($/KW)	Capacity Credit ($/KW)	Percent Capacity Credit
New England Power Pool	950	150	16
Niagra Mohawk	1000	500	50
Kansas Gas & Electric	500	140	28
Northwest Pacific Power (Coast Site)	650	250	38
Northwest Pacific Power (George Site)	500	200	40
Southern California Edison	300	80	27

Source: Edward E. Johanson, JBF Scientific Corporation, *Synthesis of WECS/Utility Integration Studies: Centralized and Dispersed,* Proceedings of the Fourth Bienneal Conference on Wind Energy Conversion Systems, October 29–31, 1979.

from the fact that by the year 2000 these utilities will have less dependency on expensive fossil fuel because of improved equipment mix and smoother diurnal-load curves. Thus the value of fuel savings relative to capacity credit will be proportionately less than it is today. This is consistent with the dominant effect of fuel savings in the NEGEA case previously cited.

In arguing against a 4300-MW coal installation, to be shared by Southern California Edison and Pacific Gas & Electric, the Environmental Defense Fund (EDF) has proposed a combination of alternative capacity additions including geothermal, cogeneration, biomass, improved end-use efficiency, and wind [19]. The combined utility-expansion plan called for 472-MW of wind power from 1980–1992. EDF argues for 2666 MW of wind. Yet no capacity credit is assumed. EDF argues, "In fact some portion of wind generating capacity can be reliably counted on for purposes of meeting peak load. Reserve margins are more than adequate, however, even if no firm wind capacity is included." This results from shifting oil- and gas-fired units to reserve status as the utility expands with non-oil-energy alternatives.

Lindley and Melton treat the issue of capacity credit for WECS on the island of Oahu, Hawaii [7]. They not only calculate the capacity credit but they also identify which particular units in the utility's expansion plan can be omitted. They state that 300 MW of wind turbines will displace 70 MW of conventional generation—capacity credit is thus 23.3 percent. Ultimately, however, they conclude, "the savings associated with these capital displacements were dwarfed by the savings associated with fuel displacement. Therefore, it was decided to completely discount the capital value of the WECS and figure the WECS solely on the basis of operational fuel savings." Wicks et al studied twenty-two scenarios related to the expansion of the New York Power Pool considering up to 18,000 MW of wind generation [23]. The work is based on hourly simulation of wind and measures each scenario in terms of the combined present value of capital, operation, and maintenance cost, including fuel, of the entire pool. Capacity credit for wind is explicitly considered, and in scenarios wind displaces baseload coal and nuclear in the expansion plan. A diversified mix of wind machines and the presence of pumped storage is shown to significantly increase capacity credit.

For the utility described in chapter 5, we have decided to assume that a wind project will have to be justified by its ability to reduce operating expenses only (mainly fuel) and that no capacity credit will be considered. While this certainly penalizes the economic evaluation of the wind project, particularly in later years, our own discussions with utility executives and planners mitigate against allocating any capacity credit to a wind-energy project in today's utility-planning climate.

Storage

The value of wind energy can be increased if storage is available to capture wind energy when it is not needed or when its marginal value is low so that it can be used later when it has a greater value (that is, shifting the wind energy captured off peak for use during peak hours). Garate evaluates several storage concepts for use in wind energy, summarized in table 3–3 [11].

The most practical decision for a utility when considering wind is to use hydroelectric storage, where it does exist, to enhance the value of operating both the wind turbines and the hydroelectric system. The project at Medicine Bow is justified entirely by the existence of hydroelectric storage [16]. Lindquist and Malver consider new, dedicated storage for a wind project at Minnesota Power and Light but conclude that the additional cost of storage for wind power alone would not be justified [2]. None of the other studies reviewed employed storage to justify the WECS project evaluated.

Since the utility chosen for analysis herein (chapter 5) does not contain any hydroelectric power, the issue of storage does not receive any quantitative analysis in this report. This choice is deliberate because the detailed hydrologic and storage simulation required to establish the value of wind power in the presence of existing storage is well beyond the scope of work that was funded.

Is should be stressed that storage systems have a value to the utility independent of wind energy. Pumped storage systems are conventionally justified on the basis that they can store energy at low cost during off-peak hours and displace high-cost generating during peak hours.

In a recent study, Winer shows that a combination of wind, pumped hydro, and conservation measures are complementary approaches to optimization of the overall utility-energy system and can result in a reduction of baseload needs [20].

Methodology of Study

Table 3-3
Value of Storage Estimates

Type of Storage	$/Kw	$/Kwh	Comments
Pumped hydro	110	6	Site limited
Lead acid batteries	75	55	Safety problems
Advanced batteries	65	30	Safety problems
Thermal concepts	150	10	Heat exchanger problems
Flywheels	200	75	Best for short term peaks
Hydrogen	200	6	Low efficiency, site limited
Compressed air	135	12	Site limited
Magnetic	—	—	Far term concept

Source: John A. Garate, General Electric Company, *Wind Energy Mission Analysis*, Department of Energy (EY-76-C-02-2578), February 1977.

Finally, a recent publication by the National Research Council reviews the state of the art of solar-energy storage [27].

Thus the issue of wind power in the presence of storage is important and should be the subject of further research.

Policy Analysis

We review briefly some of the studies that have analyzed policies relating to wind-energy investments using quantitative analysis and models. The focus here is on the analytical techniques employed.

Boyd et al considered the utility market for WECS, among other market opportunities such as residential, agricultural, and industrial [21]. The report uses a conventional revenue-requirements methodology. It presents a series of nomograms that allow the user to specify an investment-tax credit, accelerated depreciation, interest on project loan, direct-purchase subsidy and property tax elimination. The result is a reduction in revenue requirements. The authors report, "The investment tax credit and accelerated depreciation appear to be the most effective WECS economic incentives relative to the investor owned utility sector."

Lotker et al. analyze the market for utility wind investments through a detailed series of nomographs which solve for revenue requirements taking fiscal and financial policies into account [22].

Stone and Webster analyzes five categories of incentives in terms of their impact on revenue requirement and breakeven cost: grants, loans, tax instruments, parity pricing for solar energy, and indirect tax or cost advantage (for example, leasing) [10]. Accelerated depreciation, investment-tax credits, reduction of tax life, and low-interest loans are all shown to effectively reduce the breakeven cost.

Financial Analysis

While there has been extensive work analyzing the economic merits of wind and other solar-electric-energy systems, none of the work cited has studied the impact of a large investment into solar-electric energy on the financial condition of the utility itself. Lindley and Melton recognize that: "Levelized methods of analysis, however, do not show the annual cost of service or the initial impact of the early years of operation. They also ignore the utilities' cash flow problems in the first few years of the WECS accounted on a levelized basis" [7]. The authors compute the cash-flow implications of the project but do not take the next step to a true financial analysis.

Felak and others at the General Electric Company have strongly advocated a rigorous evaluation of the financial impacts of any major utility investments [24, 25]. As part of its analytical services to utilities, GE developed its own financial-simulation program for regulated utilities. This program is described in more detail later because we selected it as the financial-analysis tool herein. GE reports that it has many utility clients for financial simulation. They have indicated that financial simulation is becoming an increasingly more significant tool in utility decisions to invest in new equipment.

In searching the literature, the only significant example of use of financial simulation to evaluate the impact of a major investment into wind and other alternative generation was the adversary work of the Environmental Defense Fund [19]. EDF argued that its alternative scenario for capacity additions would eliminate the need for the 4310-MW Allen-Warner Valley coal plant, a joint project of Southern California Edison and Pacific Gas & Electric. Neglecting a capacity credit for wind, they first demonstrate that the following equipment mix can meet the same reliability criterion:

1279-geothermal

1983-MW cogeneration

2194-MW wind

340-MW biomass

565-MW end-use efficiency

Then using their own financial-simulation model, called ELFIN, EDF compares the financial impact of their plan versus the utilities' plan. The following results are displayed over a twelve-year period for each of the utilities studied:

Revenue requirements

Cash earnings/book earnings

Methodology of Study

Capital expenditures

Allowance for funds used during construction (AFDC)

Internal to total-financing ratio

Cash-interest-coverage ratio

Using financial simulation, EDF argues that both ratepayers and stockholders will benefit from the alternative expansion plan in the long run, even though the EDF scenario results in increased capital expenditures. ELFIN does not simulate the effect of regulatory lag. While financial simulation models are used extensively by the electric-utility industry, their appearance in published literature is understandably scarce. Financial projections in response to alternative business strategies are considered an internal confidential matter.

Simulation

Overview

Our approach to studying the effects of policy variation on a utility's ability or motivation to invest in solar-electric facilities, or other fuel-saving technology, is illustrated in figure 3-2, and explained here.

First, for the case study, a wind project was chosen, and a representative site for meteorological conditions was selected (see chapter 5 for descriptions). A model that simulates wind-energy production each hour was developed.

Second, a production-cost model was selected that assigns a utility's available equipment to meet the load each hour in a manner that minimizes production cost, satisfies spinning-reserve requirements and allows for machines to be put out of service for scheduled maintenance and repairs.

The inputs to the production-cost model correspond to the utility being studied. We selected a "synthetic" or "regional" utility, developed by the Electric Power Research Institute (EPRI) specifically to accommodate policy studies. The subject is an investor-owned utility.

Thus we developed a "baseline" scenario in which a major wind project was designed for a regional utility, and the annual electrical production costs were computed by simulation for each of the years 1985–1999. This simulation was also performed for the case of a no-wind project so that the savings in production cost could be compared with the investment (see chapter 5), and the impact of the project on the utility's financial condition could be assessed, as described in the following paragraphs.

The series of annual production costs without the wind project is input to the utility financial model. The output is a report of the utility's financial

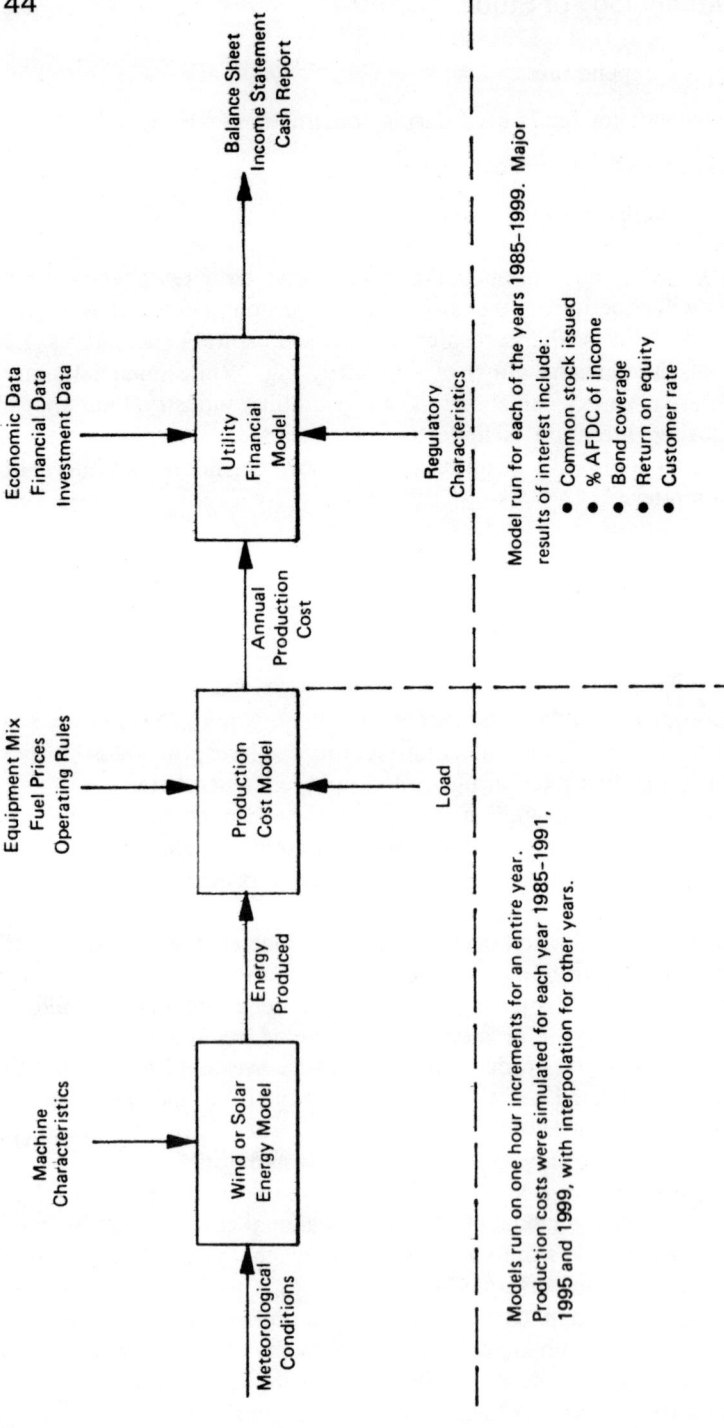

Figure 3-2. Overview of Simulation for Case Study

Methodology of Study

operating results for each of the years 1985–1999. Then the series of annual production costs, reduced by the wind project, together with modifications to "investment data" stemming from the investment in the wind project, is input to the utility financial model. This produces a second series of financial reports for the years 1985–1999. By comparing the corresponding years of the two financial reports, we can examine the effects of the wind project on the financial health of the utility (see chapter 6).

The preceding financial reports are based on a single set of assumptions about the regulatory characteristics of the utility. If the financial performance of the utility in response to the wind project is to be improved, the regulatory characteristics need to be modified. Thus for the same series of annual production costs, we successively reran the utility financial model for each year 1985–1999, each run being characterized by a different regulatory rule or condition. In this way we were able to test a variety of hypotheses related to whether a given regulatory modification would improve the financial performance of the utility in the presence of a wind project, so as to constitute an incentive for the investment.

Wind-Energy Model

The wind-energy model has the following components:

1. A tape of actual wind speeds recorded each hour for a period of one year at a representative site.
2. Empirical data on the relationship between wind speed and power production each hour for the MOD-2 wind turbine.

The choice of site for simulation of wind speed and the assumptions used are discussed in chapter 5. The resulting hourly wind speeds were at the 61-m (200 ft) hub height of the MOD-2 machine.

The relationship between wind speed and wind power is shown in figure 3–3 for two cases:

1. The output of the 1000-MW array is a simple multiple of the output of a single array.
2. The output of the 1000-MW array reflects diversity of wind over the array following data given by Justus and Hargraves [28] and Justus and Mikhail [29].

The so-called Justus-Cluster curve is a modification of the simple multiple curve determined as follows.

V_1 = wind speed from the data tape at the machine hub height

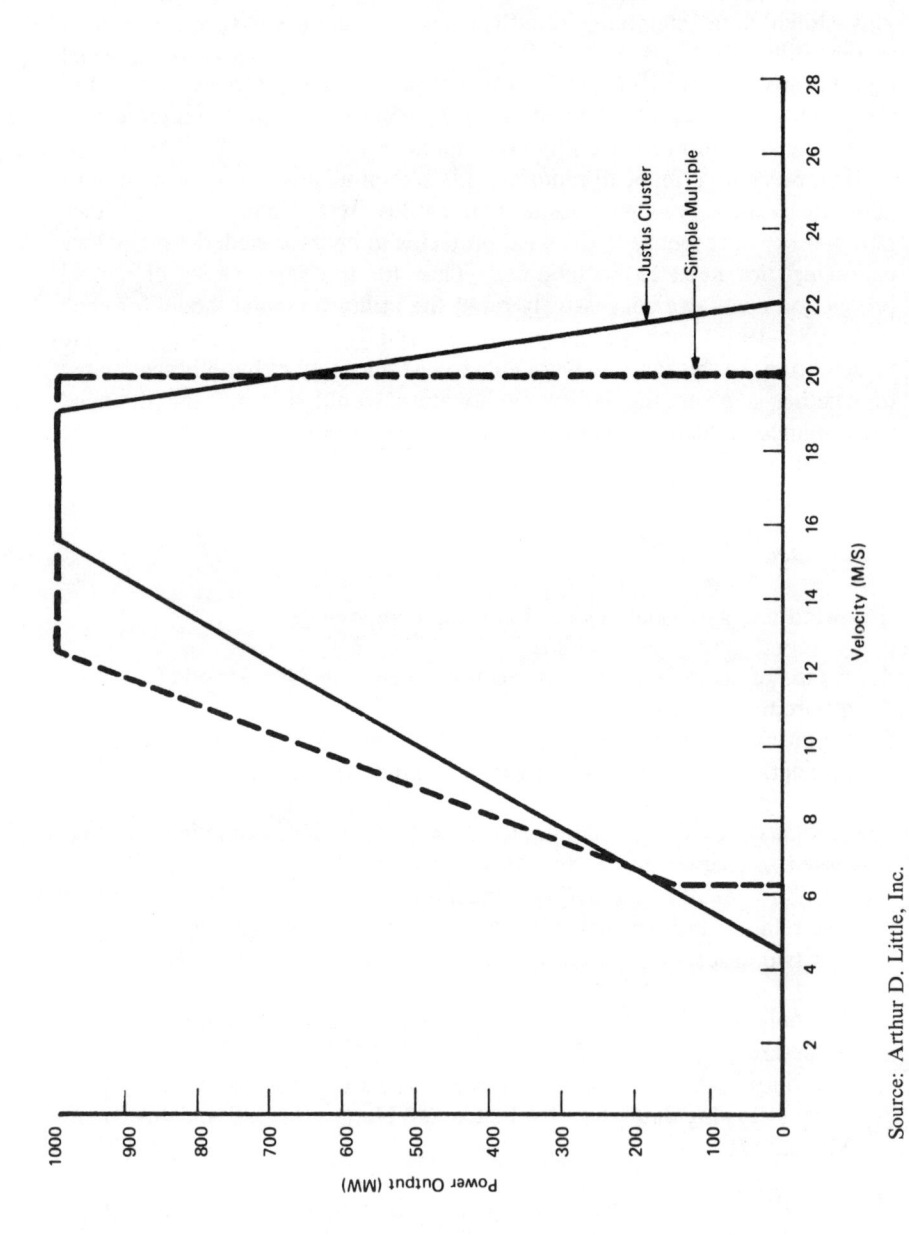

Figure 3-3. Average Hourly Wind Velocity versus Power (MOD-2 1000 MW Cluster)

Source: Arthur D. Little, Inc.

Methodology of Study

V_n = effective array wind speed for use on the simple cluster curve.

Then, depending on the size of the array, V_n is determined as:

$V_n = V_1$	up to 100-MW array
$V_n = 1.49 V_1^{.811}$	400-MW array
$V_n = 2.15 V_1^{.630}$	700-MW array
$V_n = 2.43 V_1^{.568}$	1000-MW array

The effect of accounting for diversity is to attenuate the simple multiple curve so that energy production begins at a lower wind speed, the range of velocities at which cluster-rated output is obtained is narrowed, and final-cluster cutout is achieved at a higher wind speed. In this way we provided a more realistic simulation. The results are shown in chapter 5.

Electrical Production-Cost Model

The basic production-cost model used here was developed by the General Electric Company–Electric Utility Systems Engineering Department (GE/EUSED) and is called "Monthly Production Simulation Program," or MPS. This is a proprietary program whose use is sold widely to electric utilities and others on a remote access basis via telephone lines.

Essentially the program selects from a list of available equipment, the most economic mix to meet the load each hour while maintaining a required level of spinning reserve. From hour to hour, the program moves generators up and down designated load points and starts up or shuts down equipment, taking into account data on scheduled and forced outage. Each hour the program accumulates the cost of electric production as the sum of the fuel and operating and maintenance (O&M) costs.

We ran this program for each of the years 1985–1999 in which equipment was either added or retired from the utility. (The description of the utility—its load and equipment mix including the expansion plan over the period 1985–1999—is given in chapter 5.) We then postulated a major wind-energy-development program (see chapter 5) and reran MPS to measure the reduced production cost in the presence of wind (see chapter 5). This gave us a measure of the value of fuel and O&M savings resulting from the wind project, and a measure of avoided costs. The results provided a basis for performing conventional engineering-economic analysis of the wind project and basic inputs into the financial-simulation analysis.

There are two basic options for simulating the assignment of equipment in the presence of a stochastic-energy source such as a WECS array. The first schedules equipment to run each hour assuming no wind and specifies the individual load points to assure meeting a spinning-reserve criterion. As wind energy is used to reduce total load, this method backs equipment to lower load points as appropriate, but no machine is shut down completely, assuring that spinning reserve increases as wind increases. The second option reduces the original load each hour prior to any equipment scheduling, effectively maintaining the same spinning reserve. One may expect that the first option is the more conservative because it assumes less predictability of the hour-by-hour mean wind speed; one can imagine that, in the first option, all equipment is left running (at least at some low level) because it was never absolutely certain that the wind would be blowing at all. The second, on the other hand, implicitly assumes that wind speed is relatively predictable from one hour to the next and may schedule fewer pieces of equipment.

The second option results in greater WECS-produced fuel savings because generating plants are more efficient (in a fuel-cost-per-kilowatt-hour sense) when they run at higher load points. To illustrate simply, suppose that:

The original customer load was 2000 MW during some hour.

The WECS output during this hour was 1000 MW.

There are two identical thermal units available, with minimum and maximum ratings of 500 and 1000 MW, respectively.

Option 1 would first commit both 1000-MW thermal units to run during the hour; after load reduction due to the WECS, it would run both units at their minimum 500-MW level. Option 2 would never "see" the full 2000 MW, however, and would commit only a single thermal unit to run at its maximum level of 1000 MW. Since running a thermal unit at 1000 MW costs less than double running the same unit at 500 MW, the second option produces the lower production cost in the presence of wind.

We have chosen the second option because of the large size of our simulated WECS array, which would have to be spread over an area of several hundred square kilometers. Over such an area, the effect of micro-scale turbulence, which can cause large fluctuations in hourly mean wind speed at any particular location, would be relatively smoothed out. The output of an array of several hundred individual machines is statistically much more predictable on an hourly basis than the response of only a few machines. Unfortunately, there was virtually no data with which to measure the dependence of hourly variation in array size. We could not use the routine National Weather Service ground data since these measurements are

Methodology of Study

affected by small-scale, topographically induced instabilities to a much greater extent than measurements at 200 ft, the hub height of the Boeing MOD-2. Thus our decision is based on professional judgment, although it is an issue that must ultimately be confronted by further research.

Financial-Analysis Model

The financial-analysis model used here was also developed by the General Electric Company and is called "Financial Simulation Program" (FSP). Like MPS, its use is widely sold to electric utilities and others on a remote-access basis via telephone lines. The input to FSP includes:

   Initial balance sheet

   Rules for new financing

   Regulatory rules

   Tax rules

   Equipment expansion (that is, capital-cost requirements)

   Annual operating costs

   Electricity generated and sold

The output consists of three reports:

   Cash report

   Income statement

   Balance sheet

Each year the program executes the following operations:

   Capital expenditures and plant retirements are factored into the plant account.

   Depreciation schedules are computed for tax purposes and for book purposes.

   Revenue is computed on the basis of the electricity sold in the current year, the unit cost of electricity having been set at the end of the previous year (regulatory lag).

   New financing requirements are allocated among bonds, preferred stock, common stock, and short-term debt, maintaining a specified maximum-debt ratio in total capitalization. Stocks and bonds are sold in minimum- and maximum-issue sizes. Any resulting shortage in new financing is made up by short-term debt. The utility's cumulative retained earnings is a major factor in new financing required.

The income statement and cash flow are computed, and the end-of-year cash position is determined. If cash is outside the limits, short-term financing is affected, and the end-of-year cash position is recalculated with modified financing, interest, and taxes. Upon completion, the balance sheet is computed.

The income taxes are computed taking into account all expense allowances, depreciation, and investment-tax credits.[3]

The customer rate is computed to meet all allowable revenue requirements and results in the target return on equity, or return-on-rate base for the current year. However, this rate is not effective until the following year of the simulation, effectively imitating the effect of regulatory lag.

The output is examined for its reasonableness. If the payout of common dividends is too low, the payout ratio for the year will be increased and the program rerun. In addition, or alternatively, the allowable return on equity or rate base will be increased prior to rerun. The first constitutes user simulation of a utility-management action aimed at supporting the price of common stock. The second simulates the act of emergency rate relief for the year in which the utility suffers undue financial stress.

The FSP model was first run to construct a base case of financial performance for each of the years 1985–1999. User intervention was exercised by adjusting the payout ratio or allowable return of equity for those years in which the expansion plan resulted in a less-than-acceptable return on common stock. Once the base case was constructed, no further user intervention was exercised in the simulation.

We then ran the FSP model for the case of an incremental investment into 1000 MW of wind power over a seven-year period, with the corresponding reduction in operating costs determined by MPS. This provided a case in which the financial impact of the wind project could be assessed, with no regulatory changes to the base case.

Finally, we systematically modified the regulatory rules and tax rules and reran FSP for the 1985–1999 period in order to test several hypotheses about the effects of various incentives on the financial condition of the utility. These results provided the quantitative basis for our conclusions about which incentives are effective and which are not.

The output of FSP is not fully presented herein. Reproduction of this amount of data for all of the cases run would be excessive. To render the data manageable and to display the effects of policy variation, we selected several indicators as the critical determinant's of a utility's decision to invest:

Methodology of Study

New common stock issued

Earnings per share

Bond coverage (percent)

AFDC as a percent of earnings

Return on equity

Customer rate

To further simplfy the data display, we grouped the parameters into five-year averages: 1985–1989, 1990–1994, and 1995–1999. This grouping clarifies the effects of policy variation.

Notes

1. JBF Scientific Corporation, *Northeast Regional Assessment Study for Solar Electric 1982–2000,* Department of Energy, Draft Report of February 1979, Unpublished.

2. J.W. Doane, R.P. O'Toole, P.B. Bos, P.D. Maycock, California Institute of Technology, Jet Propulsion Laboratory, *The Cost of Energy from Utility Owned Solar Electric Systems, A Required Revenue Methodology for ERDA/EPRI Evaluations*, June 1976.

3. This study was performed based on income-tax rules in effect for electric utilities in early 1981. Thus, any effects of the Economic Recovery Tax Act of 1981 were not explicitly addressed.

References

1. Doane, J.W., O'Toole, R.P., Bos, P.B., Maycock, P.D. California Institute of Technology, Jet Propulsion Laboratory. *The Cost of Energy from Utility Owned Solar Electric Systems, A Required Revenue Methodology for ERDA/EPRI Evaluations,* June 1976.
2. Lindquist, O.H., and Malver, F.S. Honeywell Inc. *The Application of Wind Power to the Service Area of the Minnesota Power and Light Company.* Department of Energy (EY-76-C-02-2618), Final Report July 1975–August 1976.
3. Asmussen, J. et al. Michigan State University. *Application of Wind Power Technology to the City of Hart, Michigan.* Department of Energy [E(11-1)-2603], December 1975.
4. Lindley, Charles A. The Aerospace Corporation. *Wind Machines for the California Aqueduct.* Department of Energy [BOA-E(04-3)] 1101 (P.A. No. 5), March 1977.

5. Johanson, Edward E. JBF Scientific Corporation. *Summary of Current Cost Estimates of Large Wind Energy Systems.* Third Wind Energy Workshop, September 19 and 21, 1977. Department of Energy [E(49-18)-2521), 1977.
6. Johanson, Edward E. JBF Scientific Corporation. *Synthesis of WECS/ Utility Integration Studies: Centralized and Dispersed. Proceedings of the Fourth Bienneal Conference on Wind Energy Conversion Systems,* October 29–31, 1979.
7. Lindley, C.A. and Melton, W.C. The Aerospace Corporation. *Electric Utility Application of Wind Energy Conversion Systems to the Island of Oahu.* Department of Energy (EX-76-C-01-2439), February 1979.
8. Marsh, W.D. General Electric Company. *Requirements Assessment of Wind Power Plants in Electric Utility Systems.* Electric Power Research Institute, January 1979.
9. JBF Scientific Corporation. *Northeast Regional Assessment Study for Solar Electric 1982–2000.* Department of Energy, Draft Report of February 1979, unpublished.
10. Stone and Webster Engineering Corporation. *Southwest Project— Resource/Institutional/Requirements Analysis.* Department of Energy, December 1979.
11. Garate, John A. General Electric Company. *Wind Energy Mission Analysis.* Department of Energy (EY-76-C-02-2578), February 1977.
12. JBF Scientific Corporation. *Wind Energy Systems Application to Regional Utilities.* Department of Energy (EX-76-C-01-2438), 1979.
13. Johanson, E.E., and Goldenblatt, M. JBF Scientific Corporation. *Wind Energy Systems Application to Regional Utilities. Proceedings of the Workshop on Economic and Operational Status of Large Scale Wind Systems.* DOE/EPRI (ER-1110-SR), March 28–30. Proceedings published July 1979.
14. Macklis, S.L. and Oplinger, J.L. General Electric Company. *Assessment of Distributed Wind Energy Systems.* Electric Power Research Institute. Reported in *Journal of the American Chemical Society,* February 1979.
15. Barret, Clifford. Bureau of Reclamation. *Wind/Hydroelectric Energy Project. Proceedings of the Fourth Biennial Conference on Wind Energy Conversion Systems.* DOE/EPRI (ER-1110-SR), October 29–31, 1979.
16. Hightower, S.J., and Watts, A.W. *A Proposed Conceptual Plan for Integration of Wind Turbine Generators with a Hydroelectric System.* Third Wind Energy Workshop, September 19–21, 1977. Department of Energy [E(49-18)-2521], 1977.
17. Higginson, R. Keith. *Report on Special Investigations Wind-Hydroelectric Energy Integration Study.* U.S. Department of Interior, Water and Power Resources Service, March 1980.

18. General Electric. Electric Utility Systems Engineering. *Monthly Production Simulation Model (MPS) for Simulating the Operation of a Generating System to Determine the Production Costs,* January 1977.
19. Environmental Defense Fund. *An Alternative to the Allen Warner Valley Energy System, A Technical and Economic Analysis,* July 1980.
20. Winer, Bette M. Arthur D. Little, Inc. *Testimony of Dr. B.M. Winer before the Federal Energy Regulatory Commission.* Project No. 2429, May 1981.
21. Boyd, D.W., Buckley, O.E., and Hass, S.M. Decision Focus, Inc. *Commercialization Analysis of Large Wind Energy Conversion Systems.* Department of Energy (DE-AC03-79ET23119), June 1980.
22. Lotker, M. et al. Booz, Allen & Hamilton. *Economic Incentives to Wind Energy Systems Commercialization.* Department of Energy (EG-77-C-01-4053), August 1978.
23. Wicks, F.E., Becker, M., and Yerazunis, S. et al. *Potential and Impacts of Wind Electric Generators upon Electric Power Systems.* Rensselaer Polytechnic Institute, IEEE [CH1523-4/80/1081], April 1980.
24. Felak, R. et al. General Electric Company. *The Necessity of Including Financial Simulation in Long Range Generation Planning.* 1978 American Power Conference, April 24 and 26, 1978.
25. Felak, R. et al. General Electric Company. *Adding Financial Simulation to Long Range Generation Planning.* 1977 American Power Conference, April 19–20, 1977.
26. Percival, D. and Harper, J. Solar Energy Research Institute. *Electric Utility Value Determination for Wind Energy,* February 1981.
27. National Research Council. Assembly of Engineers, Energy Solar Energy Panel of the Committee on Advanced Energy Storage Systems. *Energy Storage for Solar Applications.* National Academy Press, 1981.
28. Justus, C.G., and Hargraves, W.R. *Wind Energy Statistics for Large Arrays of Wind Turbines (Great Lakes and Pacific Coast Regions).* Georgia Institute of Technology, Report No. RLO/2439-78/3, 1978.
29. Justus, C.G., and Mikhail, A.S. *Energy Statistics for Large Wind Turbine Arrays.* Annual Report, May 1, 1977–April 30, 1978. Georgia Institute of Technology, Report No. RLO/2439-78/3, 1978.

4 Perspectives on the Electric-Utility Industry

This chapter provides a summary of issues of current concern to the investor-owned electric-utility industry. In evaluating incentives for investment into wind power or any other new technology, it is important to consider the overall context, which includes the following aspects:

The regulatory environment for electric utilities

Utility perceptions of risk and reward

The financial condition of the electric-utility industry

Current issues in utility regulation

Recent legislation and policies affecting utility investment in wind power

State initiatives: the California experience

This chapter introduces a series of complex, often controversial issues, all of which have relevance, directly or indirectly, to the issue of electric-utility investments into wind and other renewable and fuel-conservation technologies. While only some of the issues raised are explored further with models as in chapter 6, the discussion provides an important background for understanding regulatory change.

Any proposed incentives to motivate utility investments into new technology need to be viewed in the entire context of the peculiar nature of the industry as a monopoly closely regulated by state governments. Investments into wind or any other energy-generating facility are made in a business environment uniquely conditioned by regulatory history. Electric-utility affairs have generated a vast literature, which we cite selectively in the following sections.

The California experience is of special interest because of the particularly advanced nature of state policy with respect to wind-resource development.

The Regulatory Environment for Electric Utilities

Utilities as a Natural Monopoly

Electric utilities are regulated because they are natural monopolies. The technology inherent in the monopoly supply of electricity is more efficient and economic than if there were several suppliers competing in a region. In return for enjoying monopoly status, utilities must subject their rates and return on investment to state regulatory authority. [1]. This status also reflects their localized nature and the restricted markets for their services. Markets are limited because of the geographical proximity of the utility to its consumers. [2]

Economies of scale has meant that it has been more efficient to generate power in large conversion systems than in small ones. The utilities have been assumed to have decreasing unit costs with size over the entire extent of the market [3, p. 313]. However, recent literature suggests this classic economy-of-scale effect does not necessarily apply to all aspects of service expansion [3, p. 320]. Some of the arguments stated include:

Large units have proven to be less reliable than small ones and have experienced higher outage rates.

Large units require greater reserve margins than do small ones.

Large units have greater financial risks that detract from their generally lower unit-variable costs.

Large units require long construction lead time for siting and environmental approvals.

Increases in the cost of capital for utilities have altered economies of scale.

It is beyond our scope to explore these issues in depth. Recent policy at both the federal and state levels has raised the issue of dispersed and decentralized power as potentially more efficient than large centralized plants. Encouragement of small power producers (for example, wind turbines) under the Public Utilities Regulatory Policies Act (PURPA) and the special tax credits available to these producers—neither of which apply to electric utilities—are evidence of this. The provision of PURPA stating that qualified small power producers are exempt from state regulation presupposes that consumers may benefit from dispersed and decentralized generation and from competition among generating sources. This is a significant departure from the traditional monopoly rationale.

Of course, there are arguments that encourage large-scale wind energy and other renewable-resource development within the context of the traditional monopoly. Wind energy itself has economies of scale that a monopoly utility can more readily capture. Utilities have the technical skills and the experience in high-quality maintenance, as well as experience in selecting new equipment. Providing incentives to third-party power producers acts as a disincentive to utilities to invest and assume the risks of a new technology.

Thus, Benson argues that similar incentives should be available to the utilities [4].

Finally, there are arguments raised against a proliferation of unregulated small power producers. Since the utility will normally pay the avoided cost of purchased power (as set by the regulatory commission), any economies in third-party generation are captured as profits so that neither the utility nor its customers benefit. Hansen elaborates on these points [5, pp. 24–44].

The Rate-Making Process

The rate-making process has generally dealt with the ability of a utility to recover the costs of providing electric power while earning a "fair and reasonable" return on its investment. Under the traditional forms of state public-utility regulation, "fair return" means a target of operating revenues over and above current operating costs (such as fuel, maintenance, overhead, as well as depreciation and taxes).

The definition of *rate base, fair rate of return*, and so on, have been subject to considerable interpretation and controversy [6, pp. 5–6; 2]. Some of the criteria that are most frequently cited for use in setting rates include:

> Rates that provide a return sufficient to attract capital to the utility—that is, giving investors a return that is commensurate with the return available from other investment at comparable levels of risk

> Rates that produce sufficient revenue to cover the utility's costs and allow it to earn a reasonable rate of return

> Rates that charge each customer or class of customers in a fashion that is reasonably related to the cost of providing the service

> Rates that reward efficiency and discourage inefficiency of management

> Rates that maintain a reasonable degree of rate-level stability

For much of its history, the electric-utility industry enjoyed rapid expansion in demand (7–8 percent annually) and steady decreases in the real cost of electricity delivered as technology improved and as economies of scale were felt. Rate increases were low or negative, and the rate-making process was largely a technical matter receiving little public notice. More recently, however, utilities have experienced large increases in operating costs as fuel prices have escalated and in capital costs of new construction in response to generally high inflation. As utilities have attempted to recover these costs, utility rate-setting proceedings have become centers of controversy, drawing in active and technically prepared intervener groups into the process. The fuel-adjustment clause is one of the major issues that arise in this context.

Utilities vehemently maintain that the rate increases that have been granted have been insufficient and that the financial health of the industry has consequently suffered. (References 7 and 8 are typical examples in the current periodical literature.)

Utility Perceptions of Risk and Reward

Utility Criteria for Capital Investment

Introducing New Technology. In approaching a new technology-investment option, a utility uses different criteria from a manufacturer or venture capitalist in that the utility focuses on customer-cost minimization rather than profit maximization. Also the utility is not concerned with short-term indices such as payback period. Rather, it takes a relatively long-term view consistent with its ability to finance a project over a long period. This permits the utility to adopt projects with long lead times. To do this with a feeling of confidence requires a stable economic and financial environment and the ability to project future conditions within reasonable limits.

The utility industry is generally characterized as being more risk averse than industry generally. This is natural for an industry that must live with the consequences of an investment under regulatory scrutiny for many years. It is also expected behavior in an industry whose performance is measured more by the reliability of its dividend payments and the dependability of its service than by the magnitude of its "profits" (return on capital).

The introduction of new technology has historically required a participative effort between the utility and the vendor. Recognizing the utility's risk averseness, those who have promoted new technology in electric power (vendor or government) have established the following general precedents:

The manufacturer provides a "turn-key" contract at a fixed price.

The manufacturer provides guarantees on the performance of the installed system.

The government indemnifies the utility against loss or damage.

The government subsidizes the cost of the new technology.

The regulatory commission allows recovery of development costs in its rate setting.

In the very early and formative stages of nuclear power, for example,

Perspectives on the Electric-Utility Industry

utilities were offered generating stations on a turn-key basis in which the vendor assumed the risks of building a complete station. Turn-key steam-electric plants have not been offered in the United States for the past dozen years, although the practice still exists to some extent in offerings in U.S. vendors abroad. Today the utilities must usually bear most of the cost and technical risks associated with major new products.

The Climate for Technological Innovation. Utility pursuit and acceptance of new generating technology in the past cannot be fully explained, nor could it have been adequately predicted by economic theories about the operation of firms, including those in regulated industries.

One frequently encounters the term *statesmanship* to describe that handful of industry leaders typically possessing strong technical backgrounds who have introduced much of the innovation into this industry. They have generally done so with the conviction that the industry would ultimately benefit from their decisions. The financial, regulatory, and political climate of today does not foster such technological statesmanship. New pressures and higher sensitivity to risk may be inhibiting some of the noneconomic forces that historically have aided in the acceptance of new technology [9].

In the three decades 1939–1969, technology improvements, system growth, and the regulatory climate provided utilities with strong incentives to invest in new plants, generally benefiting from economies of scale and declining unit costs. Today there are new perceptions about power-plant economics and many new institutional concerns. As a result, utilities now have little incentive to invest in new technology, particularly intermittent-energy sources such as wind. The current adverse climate can be characterized by the following recent developments.

The effects of inflation on the cost of operations and on construction expenditures

The availability and cost of fuel for the generation of electricity

The continued inability to fully recover increased fuel costs through automatic fuel-adjustment clauses

The supply and high cost of capital

Difficulty in obtaining sufficient return-on-invested capital and in securing adequate rate increases when required

Compliance with changing and more restrictive environmental and safety regulations

Uncertainties regarding the construction and fueling of nuclear-generating units

Licensing and other delays affecting the construction of new facilities

The effects of conservation in the use of electric energy

The changing targets of new federal regulation including the National Energy Act

Utility companies have been experiencing these problems in varying degrees and are unable to predict the effect of these factors on electric-power operations or even on the operations of subsidiary companies. (For example, the February 11, 1981, preliminary prospectus of Texas Utilities Company, in connection with issuance of 5 million shares of common stock, calls attention to these problems.)

This unfavorable climate necessarily influences the criteria utilities bring to bear on major new investment decisions. In general, the result has been a reinforcement of conservative criteria more related to current financial-management needs rather than to long-term optimality. Utilities find it extremely difficult to implement desired generation-expansion plans partly because of their inability to manage the financial requirements and partly because of the unexpected strength of intervener interests that are contrary to the utility's concepts of economically optimal expansion. A reluctance to risk new technologies under these circumstances is to be expected.

Conservatism and Risk Aversion

Conservatism and risk aversion in making new investments stem in part from a tradition of reliability but also represent a response to a regulatory environment that neither rewards risk taking nor protects against the consequences of a poor risk.

Assume that a utility makes a substantial investment into a new technology. If the technology fails to meet expected performance standards, operating costs will increase but rate setting commissions may delay or completely disallow a pass-through to the customers. Thus, other things being equal, dividend-paying ability is reduced.

On the other hand, if the technology performs as well or better than expected, operating costs may decrease, but rate-setting policy would typically pass most of the benefit to the ratepayer and not to the stockholder whose money was at risk. This situation reinforces risk aversion and is not conducive to experiments with new technology at investor expense. The result is an extension of the introduction period for new generating technologies and may not be consistent with the goal of achieving the lowest feasible rates for customers [9].

Investments into wind energy, in particular, involve the following risks [10, p. 5; 11, p. 6].

Perspectives on the Electric-Utility Industry

The wind resources may prove to be lower than predicted.

The wind turbine may produce less than expected at rated speed.

Maintenance costs may exceed projections as parts wear out.

Downtime for repair may exceed projections.

Large wind turbines suffer risk of tipover and destruction in extreme winds such as hurricanes.

Unpredictable environmental problems related to electromagnetic-wave interference, noise propagation, and bird collisions can arise.

Perceived Barriers and Suggested Incentives

There is little experience to indicate system reliability, component life, and relative performance of small versus large machines and horizontal versus vertical machines [12]. A typical utility position on the need for operating experience to increase confidence in reliability and choice of technology is clearly stated in the Pacific Gas and Electric Company of California Supply Plan [13, p. 17].

> One of the major impediments to commercialization is the lack of a conclusive demonstration of reliable operation of wind turbine generators (WTGs) connected to utility grids over extended time periods. To confirm the technical reliability of WTGs will probably require several years of operation with several WTGs connected to utility grids.

The risk is increased if a utility relies on wind energy not just to displace fuel but in lieu of other generating capacity (that is, capacity credit). Since a capacity credit may be a major part of the economic benefit, this further inhibits a tendency to invest [14]. If the wind system proves to be unreliable, the utility may be forced to purchase higher-cost power from the grid [15].

If utilities are to find investments in wind turbines attractive, there is a need for financial incentives yielding greater potential benefits than are currently available through the regulatory process.
As stated by Lerner [11, p. 24]:

> The utilities are the key user in the commercialization of large wind systems, yet the benefits to the utilities of accelerating the commercialization process are not nearly as great as the perceived risks in having to absorb losses if the early investments do not succeed. If the utilities are to play a major role in early commercialization of large wind systems, they will need to see greater benefits.

The utility perspective with regard to investment in risky new technologies is evident in a recent statement by Dr. L.T. Papay on behalf of Southern California Edison [16], and associated events. In October 1980, the company announced a policy change regarding goals for accelerated development of alternative-energy sources. Edison's goals for firm resource additions for this decade were stated as:

Wind	120 MW
Geothermal	420 MW
Solar	310 MW
Fuel cells	130 MW
Cogeneration	300 MW
Hydro	620 MW

Dr. Papay's statement, however, says that the early precommercial costs for renewable resources will be higher than for conventional plants, resulting in a greater degree of risk for them. Papay's statement further suggests several policies that would help the utilities.

Put construction work in progress in the rate base

Give a higher rate of return

Assure normalization of tax credits

Provide guaranteed bonds

Allow accelerated depreciation

Allow expensing of the first 100 units

Make incentives that are now available to private investors also available to utilities

Provide relief from resource-board mandates requiring retrofits of existing plants

Institute government sharing of noncommercial R&D costs

Provide government insurance for failure

A major working hypothesis of this study has been that the perceived risks of investments into wind turbines exceed the potential benefits of a technically successful project and that this is the case even for a wind project that has an otherwise satisfactory engineering-economic evaluation (that is, high internal rate of return, short payback period, low-levelized busbar cost of energy

Perspectives on the Electric-Utility Industry

produced, and so on). This hypothesis was confirmed by the research reported in this chapter and is further verified in chapter 6.

The sources cited thus far and discussions with persons involved in utility-regulatory affairs led to the incentives chosen for further quantitative analysis in chapter 6.

The Financial Condition of the Electric-Utility Industry

We have seen that wind energy and other alternatives to conventional generating equipment impose risks that need to be carefully considered in making any investment decision. We have also seen that the regulatory environment within which the utility, as a monopoly enterprise, must conduct its business further conditions the decision processes respecting investment such as wind energy.

One of the major factors in the future of utility investments into new and innovative technologies is the financial condition of the industry itself. To the extent that financial factors affect all investments, they have a particular impact on investments perceived to be risky.

Over the last dozen years, the financial condition of the electric-utility industry has deteriorated. This is seen most graphically in figure 4–1, taken from a Standard & Poor's report that shows the decline in interest coverage from the ratio of about three times in 1968 to only slightly above two times in 1980. The latter level is often a cause for alarm in regard to the loan covenants in utility-bond-indenture agreements, which may restrict additional financing if coverage falls below 2.00 (or, worse, may consider the utility to be in a default condition, pending emergency rate relief). Not unexpectedly, the changes in electric-utility-bond ratings (number of upgradings or downgradings) has approximately mirrored the change in interest coverage, as shown in table 4–1.

The reason for this deteriorating financial condition have been treated extensively by the financial services and the media; they include the rapidly escalating costs of debt, together with the massive amount of debt sold by utilities to meet load-growth forecasts; while inflation, regulatory lag, and politics have combined to hold back rate increases and earnings growth. Moreover, the expected load growth did not materialize, due to direct conservation efforts plus the impact of the elasticity of demand for electricity as its price increased dramatically. As demand fell, the kWh base over which to spread the rising fixed costs of new-plant investment diminished so that rates had to rise even further to provide utilities a reasonable rate of return. Resistance by the public to further rate increases exacerbated the situation. Utilities had to sell more common stock to raise capital and at prices below book value as investors raised the risk premium on utility equity. The resultant dilution retarded earnings-per-share growth, making the stocks even less attractive (see figure 4–2).

Trends in Debt Ratios and Interest Coverage

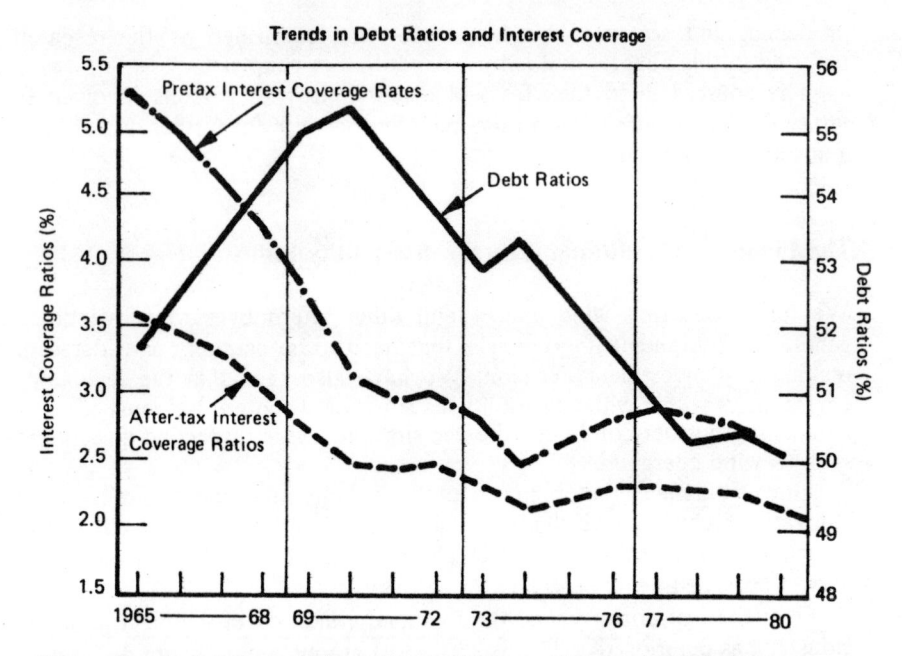

Source: *Standard & Poor's Industry Survey, Utilities—Electric,* April 17, 1980. Reprinted with permission.

Figure 4-1. Trends in Selected Indicators of Utility Financial Health

As a consequence, utility-bond ratings have fallen, as shown on table 4-1. Whereas a good growth projection was favorable for utility ratings in the sixties, "no growth" is often preferred today because of the financial strain induced by major new investments.

Table 4-2 indicates the slowdown that the electric-utility industry has experienced. It also indicates the increasing dominance of fuel costs in the total cost of electricity as the increase in cost of fuel has outpaced other costs. While capacity growth has slowed to nearly match demand growth, new

Table 4-1
Changes in S&P's Electric-Utility Bond Ratings

Year	Upgradings	Downgradings	Year	Upgradings	Downgradings
1979	3	15	1974	0	15
1978	7	9	1973	1	6
1977	9	5	1972	0	17
1976	10	8	1971	0	10
1975	5	15	1970	4	10

Source: Standard & Poor's Industry Survey, *Utilities—Electric*, April 17, 1980. Reprinted with permission.

Perspectives on the Electric-Utility Industry 65

Source: Standard & Poor's Corp. and Arthur D. Little, Inc.

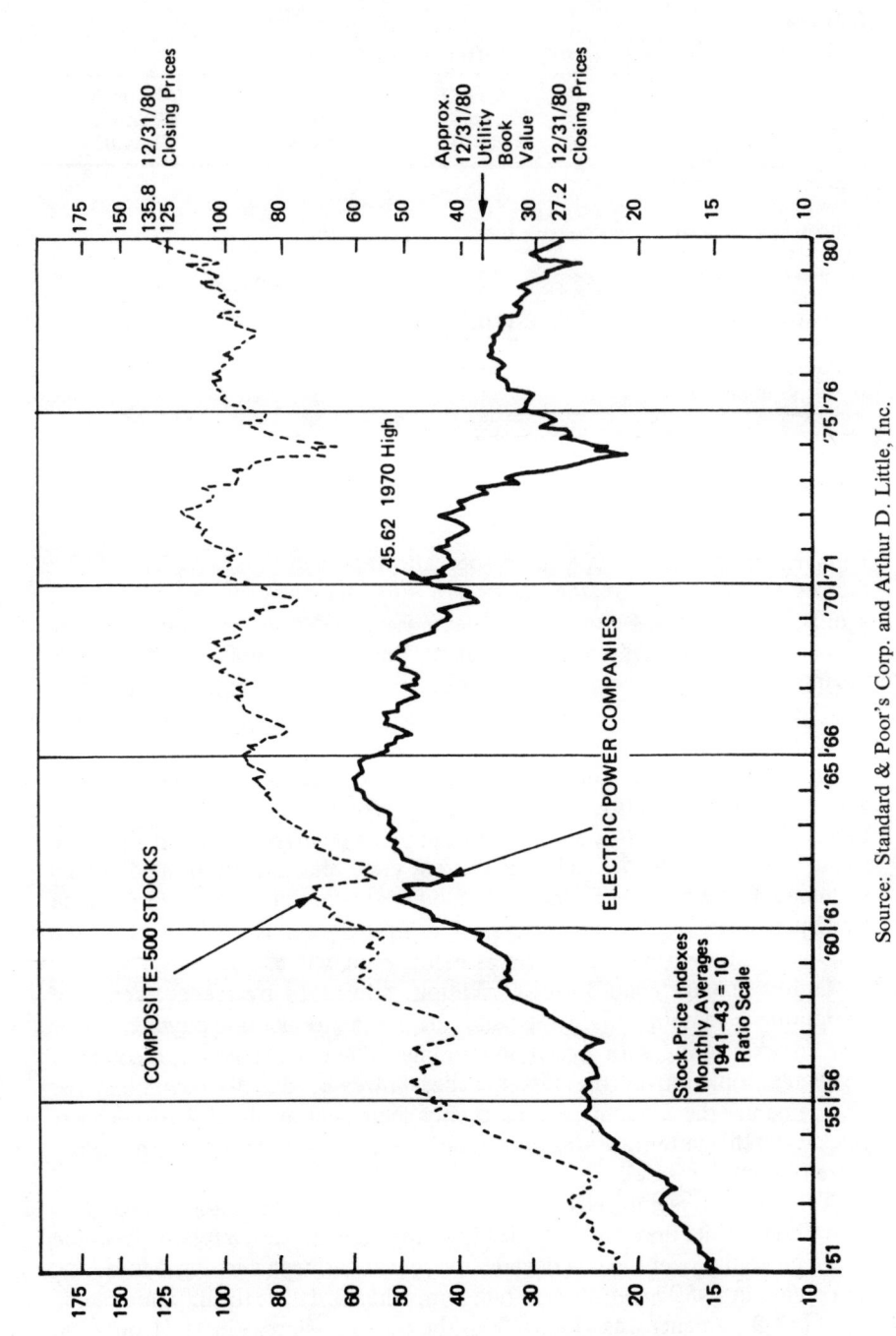

Figure 4-2. Market Action

Table 4-2
U.S. Electric-Power Industry Annual Growth Rates

	1969–1979 (10 Years) (Percent)	1974–1979 (5 Years) (Percent)	1977–1979 (2 Years) (Percent)
Capacity, kilowatt-hours	6.7	4.7	3.3
Kilowatt-hours sold	4.7	4.4	3.2
Utility revenues, dollars	15.5	14.8	11.4
Fuel cost per kilowatt-hour			
Composite, fossil	21.0	11.8	8.6
Overall, including hydro and nuclear	18.0	11.5	8.1
Electricity revenue per kilowatt-hour	9.3	10.2	7.8
Memo: Consumer price index	7.3	9.4	10.7

Source: Arthur D. Little, Inc., based on Standard & Poor's Corp. and Edison Electric Institute data.

capacity costs, on average, more per kilowatt than existing plant. These factors mean system average costs are still increasing, in contrast to the 1960s, when utilities experienced decreases, and rates actually declined. More recently utilities have been facing the difficult political situation of having their bills increase at a rate greater than the consumer price index, although table 4-2 and other indications suggest that this condition may now be stabilizing.

Investors in utility shares typically regard them as "income" stocks. There is a high correlation between utility-stock yields and interest rates [37]. Indeed, investors in utility stocks in recent years received most of their return in the form of dividends, including current yield plus growth in the dividend payment. There has been relatively little expectation of capital gains, due to inadequate rate relief, which caused a squeeze on earnings and/or the dilution resulting from selling shares below book value.

Utility managements have a tradition, reinforced by recent events, of attempting to maintain dividends, and, they hope, to increase payments over time commensurate with growth in earnings. The dollar amount of common dividends tripled over the last decade; however, due to new financing requirements, the amount per share only increased at about 3 percent per year. Even this latter rate was above the rate of growth in utility earnings per share of common stock.

Table 4-3 presents selected financial statistics for investor-owned electric utilities. Note that while utilities have managed to earn a relatively stable return on equity over the last decade, the return has been slipping. Moreover, 11 percent in 1980 means less in real terms than in 1971; the inflation rate in 1980 (12.8 percent) was greater than the return, whereas in 1971 inflation was only 3.3 percent. In this kind of climate, it is difficult for utilities to raise

Table 4-3
Selected Financial Statistics: Investor-Owned Electric Utilities

	1971	1972	1973	1974	1975	1976	1977	1978	1979[E]	1980[E]
Financing costs										
S&P AA bond yield (average), percent	7.7	7.5	7.9	8.6	8.8	8.8	9.1	9.1	10.3	11.8
Earnings-to-stock-price ratio (average), percent	8.0	8.6	10.3	13.6	13.7	12.0	11.8	12.7	14.3	15.7
Fixed-charge coverage										
After-tax interest coverage (times)	2.4	2.4	2.3	2.1	2.2	2.3	2.3	2.3	2.2	2.1
Interest coverage ex-AFUDC (times)	2.1	2.1	2.0	1.8	1.8	1.9	1.9	1.8	1.7	1.6
Return										
Return on common equity, percent	11.4	11.7	1.5	10.6	11.1	11.5	11.4	11.3	11.3	11.0
Annual EPS growth, percent	1.5	6.9	1.0	−5.5	7.4	4.1	4.8	−0.2	1.9	−2.0−
Capital structure										
Equity capital to total long-term, percent	34.6	34.6	35.1	34.4	34.7	35.6	36.5	37.2	37.3	37.3

Source: Standard & Poor's Corp. and Arthur D. Little, Inc.

capital for any kind of major investment, let alone technologies considered risky.

The simplest solution offered to this "crisis" is to allow rate increases sufficient to restore the industry to its earlier historical financial health. (See reference 38 for a discussion of the application of modern finance theory to rate regulation.) This would then free the utilities to make better and more diversified investment decisions. Renewable technologies would be included to the extent that they are economic and help minimize the cost of generating electricity. Under the circumstances, wind energy would require no particular incentives for widespread acceptance, assuming it were technically feasible and economic.

Unfortunately, rate relief to restore industry financial health is a very complex and controversial issue. As seen in the following section, there are several major issues in utility rate making that seem to assure a continuity of controversy in the foreseeable future. There is no assurance that a financially improved industry would regard wind-energy investments any more favorably. The perceptions of risk and reward discussed in the previous section will still prevail.

In chapter 6 we will focus on several scenarios of selective regulatory change whose purpose will be to improve the investment outlook for wind energy, assuming technical and economic feasibility. These will involve mechanisms for improving the financial performance of the utility as a direct consequence of the investment into wind. We assume that if the investment is perceived as financially constructive, it is more likely to be made. With this in mind, we will raise several hypotheses of regulatory change to be tested with the aid of a financial model in order to measure whether this action would improve financial performance.

Thus we study policy changes in a highly selective manner, seeking to stimulate renewable-energy investments without trying to solve the chronic financial and regulatory problems that currently characterize the industry. In so doing, however, we require at least a general understanding of some of the major issues raised in the larger context of providing the industry with general financial relief as presented in the following section.

Current Issues in Utility Regulation

There are several issues in utility regulation that are currently being debated at rate hearings and by the various state legislatures as they define the regulatory process. These include:

Automatic fuel-adjustment clauses

Construction work in progress in the rate base

Allowance for funds during construction

Averch-Johnson effect

Depreciation allowance

Normalized versus flow-through accounting

Regulatory approval requirements for projects

Regulatory lag

Each of these issues is dealt with differently on a state-by-state basis and sometimes on a project-by-project basis. A general introduction to these issues is important for understanding the debate on remedies to the adverse financial condition of the electric-utility industry and, within this context, to evaluate the financial-model results on investments into wind energy presented in chapter 6.

Automatic Fuel-Adjustment Clause

The automatic fuel-adjustment clause (AFA) is a regulatory device that allows a utility to increase (or decrease) its rates without a formal rate request in order to compensate for a change in the cost of fuel. The objective of AFA clause is to protect the utilities' allowed rate of return and overcome the adverse effects of regulatory lag caused by rapid increases in the price of fuel over which the utilities presumably have little or no control.

Most states have some form of AFA clause. These clauses are usually automatic, so a utility's increased costs for fuel can be passed on to ratepayers during the year without regulatory-process approval and the concomitant delays that would entail. Instead of formal rule making and hearings, the fuel-cost increase is usually regulated by a formula that attempts to match the time periods during which a utility incurs a fuel increase and receives the revenue to compensate for the increase.

It has been argued that AFA clauses act as a disincentive for utilities to take action (for example, to invest in renewable-resource-generating facilities) to reduce their use of high-cost fossil fuel (see reference 17a for a discussion of the theoretical and empirical arguments against fuel-adjustment clauses). If a utility automatically recovers its higher fuel costs, it has no incentive to make investments that reduce these costs. If an investment is perceived by the utilities as risky, which is the case with wind turbines, the discincentive is exacerbated.

The importance of the relationship between fuel-adjustment clauses and efficient use of fuel was recognized in Section 113 of the Public Utility Regulatory Policies Act (PURPA). It provides that an AFA clause must be evaluated by the state regulatory authority at least every four years to determine if it provides incentives for the efficient use of resources and at least every two years to ensure the maximum economies in those operations and fuel purchases that affect the rates to which the clause applies.

The guidelines proposed by the Department of Energy (DOE) [18] under the act recognize the potential disincentive that AFA clauses have for efficient fuel use.

> While fuel clauses are intended to protect a utility's earnings stability, they do not ensure that a utility's fuel purchase policies and practices are the most economical and efficient possible. In fact, fuel adjustment clauses in providing for automatic passthrough of certain costs, may permit and even foster, inefficiency in fuel procurement practices and other utility management decisions. The automatic adjustment clauses standard established by PURPA is intended to address this situation by requiring regulatory authorities to build efficiency incentives into automatic adjustment clauses.

The guideline is organized around four major issues:

Incentives for effective procurement practices

Incentives for economic and efficient generation of electricity

Incorporation of fuel-adjustment charges into the rate structure

Incentives for displacement of scarce fossil fuels in the generation mix

These guidelines are not intended to encourage use of AFA clauses. Rather, they are intended to build more efficiency incentives into their use. It should be noted that the guideline is only advisory in nature and not binding on the state regulatory authorities. It does, however, constitute DOE policy regarding consideration of a federal standard for fuel-adjustment clauses.

In developing the guidelines, DOE did consider prohibition of AFA clauses. However, they decided against this approach because it would adversely impact the cash flow of most utilities.

A policy approach to fuel-adjustment clauses, consistent with the DOE guideline, was proposed in a recent report to DOE [19]. The report proposes that the percentage of fuel-cost increase passed through be a function of the efficient use of fuels and their competitive procurement. For example, they would tie the percentage allowed to competitive procurements to fuel substitutions and to other measures of enhanced productivity.

Modification of AFA clauses could be used as in incentive to encourage investments in renewable-resource-generating facilities such as wind turbines. The continuation of an AFA clause, or the percentage increase that is allowed could be made contingent upon a specified level of investment in fuel-saving equipment (for example, wind turbines).

Construction Work in Progress/
Funds Used during Construction

A recurring problem in utility regulation is whether construction work in progress (CWIP) should be included in the rate base thereby allowing the utility to pass construction financing costs through to ratepayers before equipment is in service. Utility construction projects are financed through a combination of debt, equity (common or preferred stock), and retained earnings (internally generated funds). While a project is under construction, a utility must pay interest on the bonds and preferred-stock portion of the financing. Utility commissions differ in the extent to which they allow utilities to incorporate these costs into the rate base.

A recent survey of state public-utility commissions shows that thirty-three commissions allow some portion of CWIP in the rate base. Eleven of these usually allow 100 percent CWIP in the rate base. While CWIP policies are generally similar, they vary widely [20] in the conditions under which they are allowed [17b].

States that do not allow for CWIP generally adhere to the traditional regulatory principle that a project must be "used and useful" before it can be included in a utility's rate base [2, pp. 178–179]. For these states the alternative is the accounting concept of an "allowance for funds used during construction" (AFUDC). AFUDC is a method whereby a utility capitalizes the current costs of construction financing (that is, interest cost and the imputed opportunity cost of equity) over the future life of the project. (See reference 17c for details on provisions and formulas for computing AFUDC.) The construction and financing costs together define the cost of the facility, which is added to the rate base when the facility begins operation. Thus the financing costs are recovered from ratepayers over the life of the plant through depreciation charges. Alternatively, where CWIP is included in the rate base, the financing costs would be recovered from current ratepayers during construction, thus obviating the need for an AFUDC credit.

In accounting terms, the AFUDC account is reported as "other income" on a utility's income statement. It is, however, a noncost item with no effect on funds flows or income-tax obligations [17d].

One rationale for use of AFUDC is to defer the financing costs to a later period in order to match the costs of the construction project with the benefits generated by the project. This practice assumes that no benefits derive from the project until it is completed and in operation. It assumes that current ratepayers should not be charged the costs of debt and equity for those assets that are not currently performing a useful service [17e].

The inability of many utilities to include CWIP in the rate has recently created significant financial problems. Without it, the utility earns no return on invested funds during long construction periods but is committed to interest payments on these funds. With greater inflation and the trend toward larger plants and longer lead time, the use of AFUDC rather than allowing CWIP in the rate base has contributed to the declining financial condition for utilities [7].

AFUDC is a noncash form of current income that is accrued on the books by a utility during the construction period. It is a deferred-income account, with the actual cash deferred until the facility is in operation, and the capitalized facility construction and financing costs are in the rate base and can be depreciated. However, with inflation and long construction periods, AFUDC has been increasing significantly as a percentage of utility income. Recent data [17f, p. 18] shows that AFUDC, as a percentage of earnings available, on the average, for common utility stock reached 44.3 percent in 1979. For 1960, 1965, 1970, and 1975 respectively, the percentages were 5.9 percent, 4.2 percent, 17.8 percent, and 31.1 percent. One result is to dilute the quality of the earnings that usually results in lower stock prices.

If CWIP is allowed in the rate base, a utility is generating cash earnings to meet its financing costs. With AFUDC, earnings are overstated to the extent of the AFUDC amount since no actual cash is being earned on the construction work in progress to meet current financing costs. These issues are discussed in several articles of the *Public Utilities Fortnightly* [17f–17i].

For states that do not currently permit CWIP in the rate base, one incentive for investment into renewable resources may be to permit CWIP in the rate base only for the renewable-energy facilities, such as wind turbines, but not allow it for conventional facilities. [21, p. 5–2].

> Although the utilities desire that CWIP associated with all technologies be included in the rate base, regulatory agencies desiring to selectively enhance the attractiveness of certain technologies may choose to permit the inclusion of CWIP associated with the preferred technology in the rate base as a method of achieving their goal.

Extending the reasoning, it can be argued that states that allow CWIP in the rate for all facilities are providing a disincentive for renewable-energy investment. By earning a return on large, long-lead-time facilities before they are completed, there may be less incentive to build smaller, lower lead time facilities. This line of reasoning is discussed in a recent report by the California Energy Commission [22].

The Averch-Johnson Effect

There is a substantial body of literature on the biases that are inherent to utility investment. Perhaps the best known of this is the Averch and Johnson (A-J) effect [23]. In simple terms, the A-J effect postulates that the regulatory structure encourages utilities to undertake more capital-facility expansion than is needed to meet their service needs. Regulation is said to encourage uneconomical expansion of investment in order to increase the size of the rate base upon which the utility's return is calculated. It is not within the scope of this study to evaluate the issue of the past or present validity of the A-J effect [24]. However, the central hypothesis about the bias toward capital may be in need of reexamination in light of the recent effects of inflation, regulatory lag, and the poor financial conditions of utilities—all of which may act in varying degrees to discourage excessive capital investment by utilities. If the A-J thesis is valid and there is a true bias toward capital-intensive projects, then utilities would be encouraged to invest in solar and wind equipment that replaces fuel with capital [25], as well as investing in nuclear plants [25, pp. 313–352].

The A-J proposition assumes that utilities are facing a classic declining marginal-cost curve, and/or declining average-total-cost curve—neither of which is the case today. Also, neither is necessarily the case in the short run with wind energy. Logic now suggests that the A-J effect should operate in reverse; that is, utilities should be minimizing the scale of incremental investments and/or diversifying away from the electric-utility business. Both responses have become evident.

Depreciation for Federal Taxes

Depreciation schedules for wind turbines are not yet available from the IRS. However, the Department of the Treasury has announced that it is undertaking a study to provide an asset-depreciation guideline class, period, range, and repair-allowance percentage for property used in converting wind energy to electricity [26]. Windfarms Limited of San Francisco expects to depreciate its equipment over twenty-five years.

Accelerated depreciation is a commonly advocated incentive for motivating investments into renewable energy. (See chapter 3, references 10, 21, and 22.) The term can be understood in two ways:

1. Within a given period, it means a schedule for biasing depreciation toward the early years. For example, say $100 is to be depreciated over

five years. Straightline method yields $20 per year. Sum-of-the-years digit method yields $33, $27, $20, $13, and $7, effectively accelerating the depreciation. Most utilities currently use some form of accelerated depreciation allowed by the IRS.
2. The term can also mean the reduction of the period over which the asset is depreciated.

In chapter 6 we apply the latter concept.

Normalized versus Flow-Through Accounting

There are two methods used by regulatory commissions to account for the tax savings resulting from investment-tax credits and accelerated depreciation: the flow-through method and the normalization method. Under the flow-through method, only taxes actually paid by the utility are recognized as allowable expenses for rate-making purposes. Thus investment-tax credits and accelerated depreciation reduce the amount of tax the utility must pay, and the tax saving for that year "flows through" to ratepayers in the form of lower rates. Under the normalization approach, the tax savings are not passed through to ratepayers in the year they are incurred but instead are deferred so that they enter the rate calculation each year over the life of the plant. Thus where normalizing accounting procedures are used, current ratepayers contribute revenue to a deferred-tax account while the tax savings are being generated. This account is not included in the rate base. The account is then amortized over the life of the asset when the asset is placed in service [10j, p. 13; 27, p. II-3].

There is considerable controversy surrounding the question of whether normalized or flow-through accounting should be used, and state regulatory agencies vary in prescribing accounting procedures. Proponents of the flow-through approach argue that any savings the utility receives in taxes should be passed on to its customers. Normalization requires that current ratepayers provide revenue for taxes that were not paid and will not be paid for several years in the future. It has also been argued that in some cases in which accelerated depreciation is involved, normalization results in a utility's not having to pay the tax in the future if the utility is growing. In such cases, in the later years of an asset's life, it is replaced by a new piece of equipment and accelerated depreciation begins again. This replacement effect means that the higher depreciation taken in early years is not repaid by lower depreciation in later years [17j, p. 14].

Utilities generally prefer the normalization approach because it improves their cash flow by enabling them to use the revenue from tax savings to finance new construction, thereby reducing financing costs. Proponents of normalization also point out that flow-through accounting has a bias in favor of present utility ratepayers as opposed to future customers who would have to pay higher rates than if normalization were used.

The federal government has also become involved in prescribing accounting procedures for utilities. The Tax Reform Act of 1969 restricts the use of accelerated depreciation on a flow-through basis by providing that for tax purposes, accelerated depreciation must be normalized. The act does have an exclusion that allows a utility to flow through the tax savings to customers if it had used this procedure in 1978 or in July 1969. Similarly, the Tax Reform Act of 1971 and the Tax Reduction Act of 1975 provide that investment-tax credits for utilities can be passed through to customers in the year of realization only if the utility flowed through the benefits of accelerated depreciation as of 1971. Utilities that were prohibited from using flow-through procedures were to select one of two allowable normalizing procedures. It appears that the intent of Congress in requiring the normalization approach was to encourage utility investment in new facilities by improving their cash flow [17j, p. 14].

Regulatory Approval Requirements

An extensive time period is usually required for utilities to obtain the required regulatory approval for construction of new generating facilities. The uncertainties and delays faced by utilities have been major factors in increasing the cost of new plants. The regulatory approval period has the greatest impact on capital-intensive facilities because of the amount of money involved [21, p. 52]. Most states require environmental-impact statements and completion of an environmental-review process for construction of new power plants. An increasing number of states have enacted energy-facility-siting legislation that requires that the environmental, cost, safety, and other factors be reviewed before a certificate authorizing construction is issued. Local authorities often require approvals in the form of building permits and zoning variances prior to construction.

It has been estimated that permit approvals for a wind-energy conversion plant could take up to three years and include:

> Federal Aviation Administration approval (as a potential hazard to air navigations if over 200 ft high or if located in the vicinity of an airport).
>
> Federal Communication Commission approval for microwave and transmitting devices normally found at central-station generating facilities.
>
> Preparation of federal and state environmental reports and draft and final environmental-impact statements. There is also the possibility that

federal loans and grants may be sufficient to constitute federal action that will require an environmental-impact statement under NEPA [28, p. 32].

Preparation and review of state application for site approval.

Hearings as part of an environmental-impact statement and site-approval process.

If controversial issues arise during the period, the approval time can take longer than three years. Among the most common issues raised are those related to environmental impacts [28]. While wind energy engenders no emissions and will not pollute the air and water, large wind-cluster installations nevertheless raise a number of environmental problems that need to be managed in the planning process [29, 30, 31, 32]. Some of these are briefly indicated here:

Safety	Airplane hazard, blade failure, tower collapse
Noise	Low-frequency vibrations that propagate, local aerodynamic and mechanical noises
Electromagnetic Effects	Interference with television reception and microwave transmissions.
Aesthetics	Visual impact on landscape
Wildlife	Collision hazard, particularly in migratory flyways
Land use	Conflicts with local conservation and recreation uses, implications of obstruction-free zones.

Regulatory Lag

Regulatory lag is defined as the time required for a rate-setting commission to respond to the utility's increased operating costs. The effect depends on the frequency of rate hearings, the book basis for rate setting, and various details of the regulatory process that vary from state to state. During periods of inflation, regulatory lag virtually assures that the target rate of return on investment is never achieved.

The financial model used herein provides a simple but realistic procedure for general simulation of regulatory lag. At the end of each year the balance sheet is examined and the return on investment (ROI) is compared with the customer rate. If the ROI is below the target (allowed by the regulators), the rate is increased so that the target ROI is achieved. However, this new rate does not apply to the current year; it applies in the following year. The ROI for the new year will fail to reach the target level under the following conditions:

Fuel costs increase.

Operating, maintenance, and overhead costs increase.

Taxes increased.

Short-term interest rates increase.

Costs of construction work in progress increase.

Load growth is overestimated.

Thus in an inflationary economy, regulatory lag assures that a utility will be chasing an unachievable target of return on equity. Regulatory lag can be eliminated by setting rates in the current year, taking into account current data on inflation of allowable costs. Fuel-adjustment clauses are thus partial remedies.

Recent Legislation and Policies Affecting Utility Investment in Wind Power

Under the Carter administration, several laws were passed with the intent of easing the nation's energy problems by reducing fuel consumption and by supporting the development of renewable-energy technologies.

For the electric-utility industry, the single most significant act relating to renewable energy and conservation is the Public Utilities Regulatory Policy Act (PURPA). The salient aspects of this act with respect to issues raised in this study are described in some detail here. Other federal acts that have a bearing are also described but only briefly.

The Reagan administration has not clarified its intentions with respect to how these acts will be applied or whether current provisions and rules will be continued or reversed. We expect that new policies to be formulated will affect the climate for investment by utilities, but we are in no position to speculate further.

Public Utilities Regulatory Policies Act (PURPA)

Introduction. The Public Utilities Regulatory Policies Act (PURPA) was enacted to encourage the small-scale production of electric power using cogeneration or renewable-energy sources. PURPA, together with the tax incentives available to cogenerators and other small power producers, provides significant financial incentives for investment in wind turbines for the generation of electricity. The fact that PURPA and the tax laws specifically exclude utilities from taking advantage of these incentives has a dampening effect on direct utility investments in wind turbines.

PURPA was one of five parts of the National Energy Act of 1978. The purpose of the act is to encourage the conservation and efficient use of energy by electric utilities by encouraging production of electric power by *cogeneration* (the combined production of electricity and process heat) and small power producers. The rationale for the law, a summary of its provisions, and rules and regulations promulgated pursuant to the act are contained in reference 33. The act attempts to remove certain barriers to generation of electricity by small power producers such as:

> The risk of regulation by state utility commissions that would limit the rate of return that a small power producer could earn
>
> The possibility that utilities would not buy the electricity generated by small power producers and pay a price for it that would make small power production profitable
>
> The possibility that utilities would charge small power producers and cogenerators discriminatory rates for backup power.

The act has two major sections dealing with cogeneration and small power production: Section 201 dealing with definitions of qualified small power producers and Section 210 dealing with the obligation of utilities to purchase power from small power producers and cogenerators, at the utilities' avoided cost. FERC issued proposed regulations implementing both sections in 1979. After hearings, final rules were issued under Section 210 on February 19, 1980, and Section 201 on March 3, 1980.

Definition of Qualified Small Power Producers. In order to be a qualified small power production facility, under Section 201 of PURPA, the following requirements must be met:

> It must have a power-production capacity that together with any other facilities located at the same site as determined by the Federal Energy Regulatory Commission (FERC), is not greater than 80 MW. The commission defines "facilities located at the same site" as facilities located within 1 mile of the facility for which qualification is sought. (FERC appears to be flexible in interpreting the 1-mile requirement. At the request of Windfarms Limited, FERC recently allowed the company to site its facilities closer than 1 mile apart but not be considered at the same site. The company sought relaxation of the 1-mile provision

Perspectives on the Electric-Utility Industry 79

because the ridgelike terrain in Hawaii where the company plans to install a wind farm precludes wider spacing.)

The primary energy source of the facility must be biomass, waste, renewable resources (for example, water power, solar energy, or wind energy) or any combination thereof. The term "primary source" means that more than 50 percent of the facility's total energy input must be in these categories.

No more than 50 percent of the equity interest in the facility can be held by an electric utility or public-utility holding company or any combination thereof or their subsidiaries.

Utilities are allowed to own outright or jointly small power-production facilities, but they do not qualify for benefits under the act if they own more than 50 percent. A utility could operate a small power-production facility that it did not own and still have the facility qualify under PURPA provided the other requirements of the act are met.

A municipality or any agency or instrumentality of a state or the federal government can own a small power-production facility and still be considered qualified under PURPA.

Utility-Purchase Requirements. As a result of PURPA and the rules promulgated by FERC under PURPA, utilities are required to purchase all electric energy and capacity made available from qualified facilities with which the utility is directly or indirectly connected. A utility that receives energy or capacity from a qualified facility may, with the facility's consent, transmit the energy to another electric utility.

According to FERC interpretation, the obligation of a utility to purchase power from and sell power to a qualified facility implies an obligation to interconnect with them. However, the qualified facility is obligated to pay the interconnection costs reasonably incurred by the utility.

Exemption from Regulation. Some of the most far-reaching provisions of PURPA legislation are those that exempt qualified small power producers and cogenerators from the regulatory and legislative provisions that govern the activities of public utilities. [A qualified facility must have a capacity of no more than 30 MW to be eligible for the exemptions, except for biomass for which a facility may have a capacity of up to 80 MW and still be exempt from state and Public Utility Holding Company Act (PUHCA) regulations.] These include:

> Exemption from state laws and regulations regarding electric rates and from state financial-organization regulation.

Exemption from the Public Utility Holding Company Act (PUHCA). (This exemption results from a provision of the FERC rules that states that a qualifying facility shall not be considered as an electric-utility company as defined by the PUHCA).

Exemption from the sections of the Federal Power Act, dealing with rate and security regulations for public utilities.

Rates Based on Avoided Costs. PURPA requires that the rates for purchase of power by utilities and for the sale of backup power by utilities to small power producers and cogenerators "shall be just and reasonable and in the public interest, and shall not discriminate against the qualifying cogenerators and small power producers." The act requires that rates for purchase of power by the utilities shall not exceed the incremental cost to the electric utility of alternative electrical energy. *Incremental cost* is defined by the act as:

> the cost to the electric utility of the electric energy which, but for the purchase from such cogenerator or small power producer, such utility would generate or purchase from another source.

The FERC rules under PURPA use the concept of "avoided cost" to define the rate that electric utilities must pay to qualified small power producers. (The PURPA legislation requires only that the purchased-power price from small power producers and cogenerators does not exceed the incremental cost to the utility. FERC, in meeting PURPA's legislative mandate to promulgate rules that encourage the maximum generation of power by cogenerators and small power producers, has provided in its rules that utilities must pay a price equal to its avoided cost.) *Avoided cost* is defined as the fixed and running costs that an electric utility can avoid by obtaining energy or capacity from qualified small power producers or cogenerators. It has two components: (1) energy costs, which include costs of fuel and operation and maintenance (O&M), and (2) capacity costs—that is, capital costs associated with the facilities needed to provide the capability to deliver energy. Avoided cost is the difference between the energy and capacity costs a utility would incur if it met a specified level of demand with its own facilities and the cost for meeting the same level of demand if it purchased energy and capacity from a cogenerator or small power producer. Agreements between utilities and qualified facilities based on other than avoided cost are not precluded. Utilities are also required to sell backup power to small producers at just and reasonable rates.

PURPA requires that within one year after FERC prescribes rules under Section 210 of PURPA, each state utility-regulatory authority must implement the FERC rules. The FERC rules establish the criteria that state regulatory authorities must use in determining the avoided cost for each utility. It is the responsibility of the regulatory authority to establish the rules that regulate the rates for the purchase (and sale) of electric power between utilities and small power producers and cogenerators. The rules also impose reporting requirements on the utilities to furnish data to the regulatory authorities on the basis of which avoided costs can be computed. (In a recent case, a federal district-court judge in Mississippi has declared Titles I, III, and Section 210 of PURPA unconstitutional. The opinion held that it was beyond the authority of Congress under the Commerce Clause of the Constitution to direct the states to comply with PURPA. It is expected that FERC will appeal the decision to the U.S. Supreme Court.)

In general, most states are in the process of issuing their PURPA rules. It appears that the price for the sale of significant amounts of power to utilities by small power producers will probably be negotiated, with the utility commissions likely to become involved only if the parties cannot agree on a price. PURPA requires that standard rates be established for each utility only for 100 kW or less of power. (Conclusions based on telephone discussions with several state regulatory commissions.)

Recent Developments. As a result of PURPA, some regulatory authorities are currently attempting to have utilities under their jurisdiction contract with third parties for the purchase of power. In response to urging by the California Public Energy Commission (PUC), Southern California Edison has sent out a request for proposals (RFP) for the purchase of electricity generated by wind parks [34]. The company has received twenty-six proposals. The RFP requests proposals for a 1-MW wind park with at least five wind turbines, in two categories: (1) wind parks having small turbines—less than 150 kW each—and (2) wind parks having large turbines—greater than 150 kW each. The RFP encourages proposals suggesting a joint venture with the utility.

Private companies are already planning to take advantage of the opportunities offered by PURPA. For example, under an agreement between Windfarms Limited and the Hawaiian Electric Co., the former will install on Oahu a wind farm containing 80 MW and sell all of the electricity generated to the facility.

The provisions of PURPA, along with the tax incentives in other legislation, provide financial incentives that offer a unique investment opportunity for small power production. PURPA gurantees the small power producer a market for the power it can produce and guarantees a price standard (that is, avoided cost to the utility) that is very favorable. To the extent that the small power producer's cost of generating power is lower than the utility's avoided cost, the profit potential of small power production is formidable—and their rate of return on investment is not subject to state regulatory authority. The most profitable situations will likely be: (1) where small power producers sell power to utilities that burn oil—which results in the highest avoided cost—and (2) where there is no excess generating capacity so that avoided cost can have both a capacity and energy component. It should be noted, however, that over time avoided cost may tend to decrease as capacity of small power production becomes a larger fraction of a utility's total generating mix and lower-cost energy is displaced at the margin.

Effects on Utilities. As a result of PURPA, the electric utilities are motivated to assume a passive role in renewable-resources development. The utilities are required to purchase the power generated by wind turbines by others and are not given the incentive to invest in their own. The policy choices embodied by PURPA are to some extent based on the assumptions that:

> Small dispersed generating facilities may produce power more cheaply than large ones.
>
> Small power producers would be more likely to invest in renewable-resource generating technology than would the utilities, given the incentives to take the risk.
>
> Utilities would be unwilling to invest in renewable-resource generating technology until the technology is tested; ownership and operation by small power producers offers the opportunity for such testing.

These assumptions, and the corresponding lack of incentives to the utility industry, are strongly contested [5].

Powerplant and Industrial Fuel Use Act

The Powerplant and Industrial Fuel Use Act of 1978 may be relevant to utility investment in renewable resources insofar as it restricts the future use of oil and natural gas by utilities. The act prohibits use of oil or natural gas as a primary fuel in any new power plant and phases out their use in existing power plants. The major provisions of the act are:

> Natural gas or petroleum shall not be used as a primary energy source in any new electric-power plant.
>
> No new electric-power plant may be constructed without the capability to use coal or any other alternate fuel as a primary energy source.
>
> Natural gas shall not be used as a primary energy source in an existing electric-power plant on or after January 1, 1990, unless such power plant

used natural gas as a primary energy source at any time during calendar year 1977.

The Secretary of Energy may prohibit the use of petroleum or natural gas in a power plant as a primary energy source if the plant has the technical capability to use coal or another alternate fuel as a primary energy source. The rules promulgated under the act do provide some exemptions for cogenerators.

Wind-Energy Systems Act of 1980

The objective of the Wind Energy Act is to establish an accelerated wind-energy-research development and demonstration program over the next ten years in the Department of Energy and to provide comprehensive programming and planning to meet wind-energy goals. The goals established are: (1) to reduce the average cost of electricity produced by wind-energy systems and (2) to reach a total megawatt capacity in the United States from wind-energy systems by the end of fiscal year 1988, of at least 800 MW, of which at least 100 MW are provided by small wind systems.

To achieve these objectives, the act directs the Secretary of Energy to establish a six-year small wind-energy-system research, development, and demonstration (RD+D) programs and an eight-year large wind-energy-system RD+D program. It also authorizes the secretary to provide financial assistance to any public or private entity wishing to purchase a small or large wind-energy system. Participants in the small wind-demonstration program would be ineligible for any additional energy tax credit for the same project. *Public and investor-owned utilities are eligible for subsidies.* Federal subsidies end by September 30, 1988, or when the secretary determines that wind-energy systems are competitive with conventional energy sources—which ever comes first.

The act also directs the secretary to initiate a national wind-resource-assessment program. This program will entail validating existing assessments of known wind resources, initiating a wind-prospecting program, and establishing a wind-data center. The act authorizes the appropriation of $1 million for fiscal year 1981.

Energy Tax Act of 1978

The Energy Tax Act grants a 10-percent business-investment tax credit for "energy property," which includes *inter alia* solar- and wind-energy property. The credit is in addition to the regular 10-percent investment-tax credit available under the Internal Revenue Code. The regular investment-tax credit does not apply to buildings and their structural components. The terms "solar" or "wind-energy property" include any equipment that uses solar or wind energy to generate electricity but *does not include property that is owned or leased by a utility*. In the case of property that is financed in whole or in part from the proceeds of an industrial-development bond, the interest on which is tax exempt, the credit is 5 percent.

Crude Oil Windfall Profit Tax Act of 1980

The Crude Oil Windfall Profit Tax Act increases and extend the credits provided in the Energy Tax Act. It increases the energy-tax credit for solar and wind (and geothermal) property from 10 to 15 percent. It also extends the expiration date of the credit from December 31, 1982, to December 31, 1985. *Solar- or wind-energy property owned by a utility is specifically excluded* from the credit. The act also has a "double-dipping" ban, which reduces the credit if the property is financed in whole or in part by the proceeds of tax-exempt industrial bonds or subsidized energy financing.

There is some ambiguity in regard to the relationship between the provisions of PURPA and the energy-tax credits available under the Windfall Profits Tax Act. *Public-utility property is excluded from energy-tax credits*. Thus, if a qualified small power producer under PURPA were to be classified as a public utility, as defined in the Internal Revenue Code, it would not be eligible for the energy-tax credit.

As was the case for PURPA, exclusion of utilities is a disincentive for their investment in renewable-resource generation.

State Initiatives: The California Experience

Some state governments are beginning to establish policies and goals for the generation of electricity using wind turbines. California appears to be in the forefront of this activity. The legislature has actually established wind-energy goals for the state: 10 percent of California's electricity could be supplied cost effectively by the year 2000. The California Energy Commission (CEC) has recommended that wind energy be given high priority in the R&D budget for meeting the state's electricity demand for the year 2000. They also recommended that utilities be directed to establish and meet goals for deployment of alternatives such as wind and that the Public Utility Commission (PUC) adopt a set of specific incentives to promote preferred energy sources such as wind [36].

Recent CEC testimony recommends that the PUC should set capacity quotas for each utility based on the CEC's goal of at least 250 MW of installed wind capacity by 1985, 1000 MW by 1990, 3000 MW by 1995, and 7800 MW by the year 2000. The goals, however, are subject to revision by the CEC as it completes its *Final Electricity Report* as a part of the 1981 biannual report. In order to promote early utility experience with wind equipment, the testimony recommends that the PUC

> Require each utility to prepare detailed wind-development plans to meet statewide wind goals and utility quotas.
>
> Permit utilities to expense $50 million of the cost of their first wind machines and provide an additional $50 million where stockholders assume any undue risk.
>
> Set up balancing accounts to cover excessive O+M costs on early machines.
>
> Review progress of utility development plans at future rate hearings.

The testimony also recommends that the PUC aid small power producers by standardizing and streamlining utility sales agreements.

It should be noted that prior to the enactment of PURPA in 1978, the California PUC had been considering ways to deregulate electricity generation and create a market for small-scale alternative energy production [5, p. 20].

The California PUC has been aggressive in pursuing its goal of encouraging utility purchase of power from small power producers. For example, it recently levied a rate-of-return penalty against Pacific Gas and Electric (PG&E) (costing them over $7.2 million) for failure to make reasonable efforts to promote cogeneration [36, p. 95].

The CEC has also proposed that $30 million be provided by the state to assist in the financing of utility purchases of wind turbines. Ten million dollars would be for utility projects undertaken in cooperation with the CEC. It would supplement investments for a joint project funded by several California utilities. The objective of the project would be to provide experience with wind turbines for both utilities and the Energy Commission. In addition, $20 million is proposed as loans to small power producers as premium payments above avoided costs in early years. The objective is to levelize revenues between early years and later years when avoided costs rise because of higher oil prices. When avoided costs rise above project costs, repayments would have to be made on the loan [5, p. 26].

The major utilities in California have responded to the policy initiative by including wind generation in their plans. Pacific Gas and Electric (PG&E) has stated that it believes that wind electric generation is the most promising of the solar-electric technologies to become commercially competitive in the near future. Consequently, they are pursuing "an active resource evaluation and development program." Their current resource plan includes 2.5 MW of wind by the end of 1982 (demonstration project). It then allows for the collection of one year of performance data before ordering the next four units. Their forecast scenario for wind is as follows:

1982	2.5 MW
1985	10 MW
1988	10 MW
1989	20 MW
1990	40 MW

Depending on the cost and results of the demonstration project, this resource development may be accelerated or delayed. Southern California Edison's (SCE) goals are for wind capacity about four times as great, but gradual development is expected until the late 1980s, after technology testing has been completed [5, p. 53]. SCE has proposed a program leaning toward the installation of 240 MW (peak rating) by 1992.

Southern California Edison has stated, however, that it faces serious problems in accelerating the deployment of renewable/alternative energy resources including:

Problems raising capital

The technological and financial risk

The need to divert capital to the conversion of oil-fired plants to coal as required by the Fuel Use Act

The higher cost of generating power from renewable/alternative energy sources [16].

The company has proposed several policy measures to lessen the effects of these constraints, as described earlier in this chapter.

In recent utility supply-plan submittals to the Energy Commission, private utilities in California are proposing 472 MW of installed wind-generating capacity by 1992. This amount, however, is significantly less than the 1500 MW that the Energy Commission believes "reasonably could be achieved" [36]. California has taken steps to make wind-turbine investment financially attractive to utilities. One policy is to allow a higher rate of return on this form of investment. Section 454 of the California Public Utilities Code provides that:

> Upon a showing by a public utility before the Commission that it has invested in projects designed to generate or produce energy from renewable resources, including but not limited to solar energy, geothermal energy (and wind sources) . . . , the Commission may allow a return on such investment established by adding an increment of from 1/2 percent ot 1 percent to the rate of return permitted on the utility's other investment.

Table 4-4
Dispersed Supply Additions: Targets Established by the California Energy Commission and Utility Plans

	LADWP[a]	Pacific Gas & Electric	Southern California Edison	San Diego Gas & Light	Total	Energy Commission[c]
Cogeneration	100	1033	598[e]	36[f]	1767	2200–2800
Biomass	0	127	0	0	127	520
Fuel cells	0	0	26	0	26	850
Geothermal	180	1310	170[g]	220[g,h]	1871	2770
Small hydroelectric	16	76	0	0	92	306
Solar	0	0	0	0	0	500
Wind[i]	0	223	249	0	472	1500
Totals	296	2760	1043	256	4355	8646–9246

Source: California Energy Commission, Energy Service Corporations: *Opportunities for California Utilities*, November 1980.

[a]Los Angeles Department of Water & Power.
[b]Submitted to Energy Commission as part of third Biennial Report proceedings, summer 1980.
[c]*Renewables and Alternative Technologies Synopsis*, September 1980.
[d]Includes Sacramento Municipal Utility District.
[e]Firm capacity is 116 MW.
[f]Identified potential that is not included in the resource plan.
[g]Does not include purchase from Mexico.
[h]Includes 70-MW purchase from Magma Power Company.
[i]Installed capacity.

There is one major qualification: To obtain the higher return, the capital cost of the new technology when added to the operating and maintenance cost must result in a lower cost per unit of energy generated or produced over the life of the system than would result from the use of conventional power-generating facilities. Thus, to obtain the higher return, the new technology must produce power at a lower cost than conventional technology. To date, utilities in California have not requested the higher rate of return allowed under the provision. From the utility perspective, this incentive "is too small to provide the stimulus needed to promote renewable/alternative energy resources" [16].

Recent legislation enacted in California contains tax incentives for renewable resource of electricity. Assembly Bill No. 2893 enacted in August 1980 allows for twelve-month or six-month amortization of the cost of alternative-energy equipment. California already has a 25-percent state tax credit for wind-equipment investment, but this credit is not available to utilities.

California has actively pursued several forms of energy alternatives for electric utilities. The Energy Commission set goals for each of seven technologies including wind. The five major utilities in the state have also established planning targets covering these technologies, as shown in table 4–4. Wind energy at 472 MW targeted by the utilities is third behind geothermal and cogeneration in magnitude.

References

1. Kahn, Alfred. *The Economics of Regulation: Principles and Institutions,* vol. 1. New York: Wiley, 1970.
2. Bonbright, James C. *Principles of Public Utility Rates.* New York: Columbia University Press, 1969.
3. Kahn, Edward. "The Compatibility of Wind and Solar Technology with Conventional Energy Systems." *Annual Review of Energy,* 1979, pp. 313, 320.
4. Benson, C.C. *Legislative, Regulatory and Institutional Barriers to Electric Utility Participation in Energy Conservation and Renewable Resources Development Programs.* Dallas Power & Light, August 13, 1980.
5. Hansen, David Bennett. *Wind Energy Development: A Policy Assessment for the California Public Utility Commission,* December 1980.
6. Mitre Corporation. *Energy Rate Initiatives: Study of the Interface between Solar and Wind Energy Systems and Electric Utilities,* March 1977.
7. Emshwiller, John R. "Big Financial Problems Hit Electric Utilities: Bankruptcies Feared." *Wall Street Journal,* January 2, 1981.

8. Utilities Lose Power on Wall Street." *Science*, January 30, 1981.
9. Smith, D.W., and Korn, D.H. Arthur D. Little, Inc. Report to National Science Foundation, *Electric Utilities and Equipment Manufacturers: Factors in Acceptance of Advanced Energy Conversion Technology*, September 1975.
10. *Ratepayer Financing of Solar Thermal Electric Commercial Demonstration Projects.* International Research and Technology Corp., McLean, VA., October 1980.
11. Lerner, James I. *Assessment of Large Scale Wind System Technology and Prospects for Commercial Application.* Prepared for the National Science Foundation, September 1980.
12. Fung, K.T., Scheffler, R.L., and Stolpe, J. Southern California Edison Co., *Wind Energy, A Utility Perspective.* Paper presented at the IEEE Power Engineering Society Summer Meeting, Minneapolis, Minn., July 13–18, 1980.
13. *Electric Supply Planning Process and Current Supply Plan.* Pacific Gas and Electric Co., San Francisco, Calif., October 1979.
14. Decision Focus, Incorporated. *Commercialization Analyses of Large Wind Energy Conversion Systems.* Palo Alto, Calif., June 1980.
15. Johanson, Edward E. *Synthesis of WECS/Utility Integration Studies: Centralized and Dispersed.* Proceedings of the Fourth Biennial Conference on Wind, October 24–31, 1979.
16. *Prepared Testimony of Dr. Lawrence T. Papay*, before the Public Utilities Commission of the State of California, Southern California Edison Co., December 1980.
17. *Public Utilities Forthnightly:*
    a. Schiffle, D. "Electric Utility Regulation: An Overview of Fuel Adjustment Clauses," June 19, 1975.
    b. Muhs, W.F., and Schauer, D.A. "State Regulatory Practices with Construction Work in Progress: A Summary," March 27, 1980.
    c. Johnson, J.R. "Losses on Investments in Construction Work in Progress," September 14, 1978.
    d. Rakes, K., and Bogortun, W.C. "AFUDC: An Overlooked Impact on Balance Sheets and Some Implications for Financial Reporting," June 5, 1980.
    e. Bunch, C.R. "The Top Effects of AFUDC: Financial Accounting Aspects," August 14, 1981.
    f. Lerner, E.M., and Breen, W.J. "The Changing Significance of AFUDC for Public Utilities," January 1, 1981.
    g. Trout, R.R. "A Rationale for Preferring Construction Work in Progress in the Rate Base," May 10, 1979.
    h. Fitzpatrick, D., and Stitzel, T. "Capitalization on Allowance for Funds Used during Construction: The Impact on Earnings Quality," January 18, 1980.

i. Hyde, T.W., Jr. "A Compromise on Construction Work in Progress Would Benefit Consumers and Investors," August 18, 1977.
j. Dahl, Alvin J. "The California Case: Controversy over Normalization," December 20, 1979.
18. *Federal Register*, vol. 45, no. 86, part V, May 1, 1980.
19. Technology and Economics, Inc. *The Efficient Supply of Electricity: An Introduction to Policy Alternatives*, September 1980.
20. Comptroller General of the United States. *Construction Work in Progress Issue Needs Improved Regulatory Response for Utilities and Consumers.* General Accounting Office, June 1980.
21. Arizona Power Authority, et al. *Southwest Project: Resource/Institutional Requirements Analysis*, volume 4. Institutional Studies, Prepared for U.S. Department of Energy, December 1979.
22. California Energy Commission. *Energy Service Corporations: Opportunities for California Utilities* (draft), November 1980.
23. Averch, H., and Johnson, L. "Behavior of the Firm under Regulatory Constraint." *American Economic Review*, December 1962.
24. Corey Gordon. "The Averch and Johnson Proposition: A Critical Analysis." *The Bell Journal of Economics and Management Science.*
25. Kahn, Edward H. *American Economic Review.*
26. *Federal Register*, vol. 46, no. 9, January 14, 1981.
27. ICF, Incorporated. *Utility Financial and Rate Impacts of Reconverting Coal Capacity to Replace Oil Generation: The Case of Northeast Utilities.* Submitted to the Department of Energy, March 1980.
28. Phillips, Paul D. "NEPA and Alternative Energy: Wind as a Case Study." *Solar Law Register,* May–June 1979.
29. Howell, W.E. *Environmental Impact of Large Windpower Farms.* U.S. Bureau of Reclamation, Denver, Colo., 1978.
30. Rogers, S.E., et al. *Evaluation of the Potential Environmental Effects of Wind Energy System Development.* Interim Final Report, ERDA/NSF/07378-75 11.
31. Battelle Columbus Laboratories. *Solar, Geothermal, Electric and Storage Systems Programs Document,* August 1977, FY 1979, DOE/ET-0041 (78).
32. U.S. Department of Energy, March 1978; California Energy Commission, *Impact of Large Wind Energy Systems in California*, June 1980.
33. Federal Energy Regulatory Commission. *Rulemakings on Cogeneration and Small Power Production*, June 1980.
34. Southern California Edison Company. *A Notice Inviting Expressions of Interest for the Supply of Electrical Energy Produced by Wind Parks*, October 31, 1980.
35. Public Resources Code, Section 25611, State of California.
36. Testimony of Vincent L. Schwent and E. Ross Deter on "Wind Energy in California," California Energy Commission, November 20, 1980.

37. Haugen, R.A., Stronyny, A.L., and Wichern, D.W. "Rate Regulation, Capital Structure, and the Sharing of Interest Rate Risk in the Electric Utility Industry." *Journal of Finance* 33, no. 3 (June 1978).
38. Robichek, A.A., "Regulation and Modern Finance Theory." *Journal of Finance* 33, no. 3 (June 1978).

5 Constructing a Utility Case Study

This chapter describes the components of a "base case," which was created to study the effects of wind energy on a typical electrical utility. These include:

The wind resource: Wind data from the northeast United States was obtained and adapted to apply at sites we judged typical for wind projects. We selected a class 4 site, which averages 7.5 m/s annually.

The wind project: We selected 1000 MW of MOD-2 wind turbines, to be installed over an eight-year period.

The synthetic utility: We selected a utility with an equipment mix and load shape characteristic of the northeast United States and estimated its expansion plan (without wind) over a fifteen-year period. We include an estimate of future fuel prices assumed for the base case.

Simulation of the utility-operating characteristics: We ran the MPS model, simulating the equipment dispatched to meet the load each hour of the year without the wind project and with the wind project assuming class 4 sites and then class 6 sites. The result is annual production costs in each case.

Financial and regulatory characteristics: For the synthetic utility we assumed a set of financial and regulatory characteristics using an actual (but unidentified) utility's published data.

Conventional engineering-economic analysis: The wind investment is analyzed in terms of the breakeven cost of the initial 100 MW, to establish the conditions for general feasibility.

This base case provides a realistic setting for studying the financial impacts of:

1. The wind project with no changes in the regulatory and policy environments
2. Various scenarios of regulatory and policy change, which follow in chapter 6.

The Wind Resource

Regional Wind Overview

Focusing on the northeast United States, an overview of the wind resource is provided by Battelle Pacific Northwest Laboratory—*Wind Energy Resource Atlas* [1]. Figure 5–1 displays average annual wind power as described in table 5–1. Table 5–2 shows the areal distribution of wind-power classes. Additional insight is provided by looking at seasonal-average wind power, as shown for Pennsylvania in figure 5–2. For example, in three seasons the ridges of the central-north are class 5 or 6, with a period of very low wind energy in the summer.

The *Battelle Wind Atlas* is not a substitute for accurate on-site measurements—a prerequisite for planning any major wind project, just as hydrologic records are essential for hydroelectric planning.

Some citations from the atlas, in addition to the maps and tables, are pertinent:

> These estimates are considered lower limits for exposed ridge crests and mountain summits. Local terrain features in these mountains can enhance power considerably.
>
> Those features indicative of high mean wind speed are:
>   Gaps, passes, and gorges in areas of frequent strong-pressure gradients
>   Long valleys extending down from the mountain ranges
>   High-elevation plains and plateaus
>   Plains and valleys with persistent strong downslope winds associated with strong-pressure gradients
>   Exposed ridges and mountain summits in areas of strong upper-air winds
>   Exposed coastal sites in areas of:
>   1. Strong upper-air winds
>   2. Strong thermal/pressure gradients
>
> Features that signal rather low mean wind speeds are:
>   Valleys perpendicular to the prevailing winds aloft
>   Sheltered basins
>   Short and/or narrow valleys and canyons
>   Areas of high surface roughness, for example, forested, hilly terrain
>
> The wind-power density shown on the maps in the atlas will not be representative of poorly exposed locations. Estimates for ridge crests and summits (the shaded areas on the maps) are lower limits to the wind

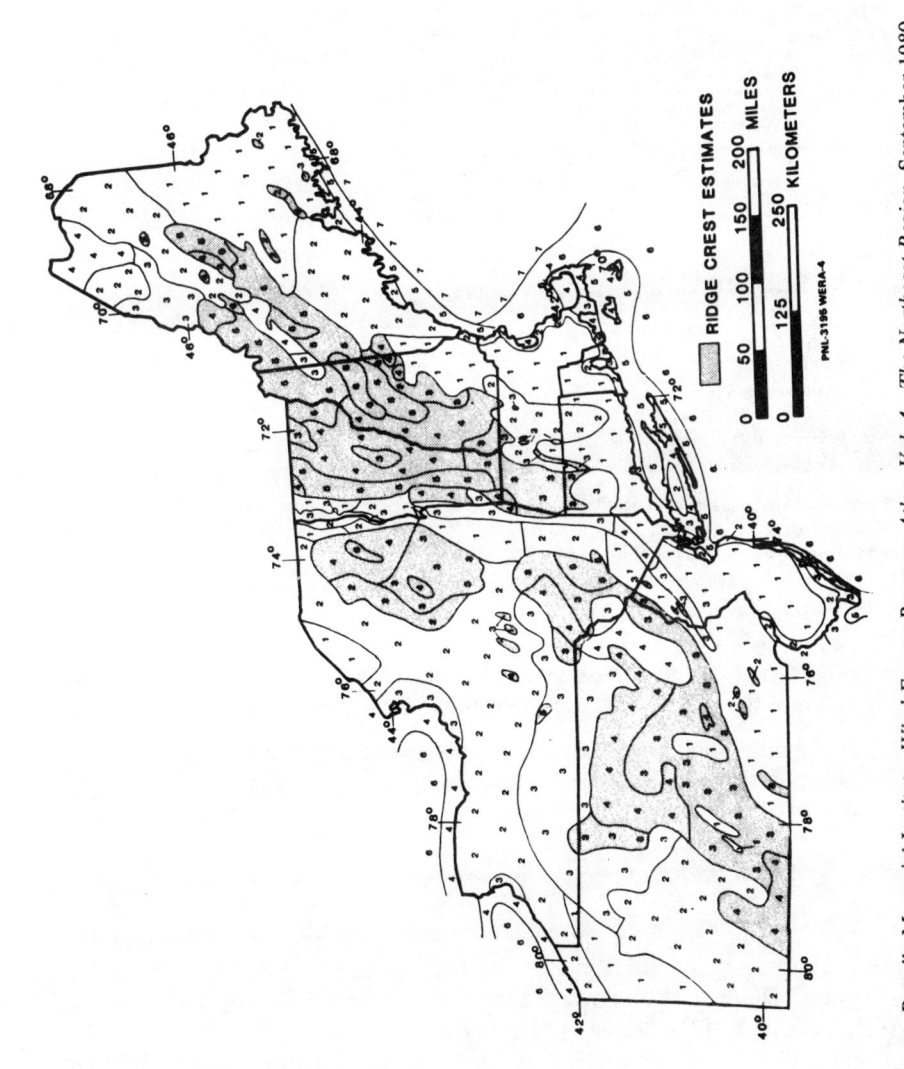

Source: Battelle Memorial Institute, *Wind Energy Resource Atlas, Vol. 4—The Northeast Region*, September 1980.

Figure 5–1. Annual Average Wind Power in the Northeast

Table 5-1
Classes of Wind-Power Density at 10 m and 50 m

	10 m (33 ft)		50 m (164 ft)	
Wind-Power Class	Wind-Power Density (W/m^2)	Speed[a] [m/s(mph)]	Wind-Power Density (W/m)	Speed[a] [m/s(mph)]
1	0	0	0	0
2	100	4.4(9.8)	200	5.6(12.5)
3	150	5.1(11.5)	300	6.4(14.3)
4	200	5.6(12.5)	400	7.0(15.7)
5	250	6.0(13.4)	500	7.5(16.8)
6	300	6.4(14.3)	600	8.0(17.9)
7	400	7.0(15.7)	800	8.8(19.7)
	1000	9.4(21.1)	2000	11.9(26.6)

Source: Battelle Memorial Institute, *Wind Energy Resource Atlas, Vol. 4—The Northeast Region*, September 1980.
Note: Vertical extrapolation of wind speed based on the 1/7 power law.

[a]Mean wind speed is based on Rayleigh speed distribution of equivalent mean wind power density. Wind speed is for standard sealevel conditions. To maintain the same power density, speed increases 5%/5000 ft (3%/1000 m) of elevation.

power expected at exposed sites. In such areas, local terrain features can enhance the wind-power considerably (for example, by a factor of 2 or 3).

As an additional complication, some land-surface forms, such as isolated hills that rise above a nearly flat landscape, may even experience a higher power density than the map indicates.

Specific Site Simulation

In order to simulate the evaluation of specific sites for the synthetic utility, we acquired a magnetic tape from Battelle containing hourly wind data for the period October 1978 through December 1979 for a hilltop site in Boone, North Carolina [2]. Boone is the site of a DOE-MOD 1 demonstration machine with 2.0 MW rated power. Unfortunately, no comparable data (that is, ridge) was available within the northeast region of interest. The Boone data, however, represents an Appalachian mountain site in northern North Carolina, near the Tennessee line, and lies in a range contiguous with the mountains of central Pennsylvania and New York. Its diurnal and seasonal variation is typical of expectations within this range of eastern mountain

Table 5-2
Areal Distribution of Wind-Power Classes in the Northeast

Power Class	Northeast	Conn., Mass., R.I.	Maine	N.H., Vt.	N.J.	N.Y.	Pa.
Land area (km²) equal or exceeding power class							
1	430,000	37,000	84,000	50,000	20,000	130,000	120,000
2	120,000	6,900	26,000	5,500	4,200	59,000	20,000
3	44,000	2,800	5,200	3,700	2,600	16,000	14,000
4	16,000	970	3,000	2,900	88	4,800	3,900
5	3,300	330	1,300	1,200	0	420	0
6	460	7	360	62	0	24	0
7	0	0	0	0	0	0	0
Percentage land area equal or exceeding power class							
1	100.	100.	100.	100.	100.	100.	100.
2	28.	19.	31.	11.	21.	47.	17.
3	1.	7.4	6.2	7.4	13.	13.	12.
4	3.6	2.6	3.5	5.8	0.44	3.8	3.3
5	0.76	0.88	1.6	2.5	0.00	0.33	0.00
6	0.11	0.02	0.43	0.12	0.00	0.02	0.00
7	0.00	0.00	0.00	0.00	0.00	0.00	0.00

Source: Battelle Memorial Institute, *Wind Energy Resource Atlas, Vol. 4—The Northeast Region*, September 1980.

WINTER

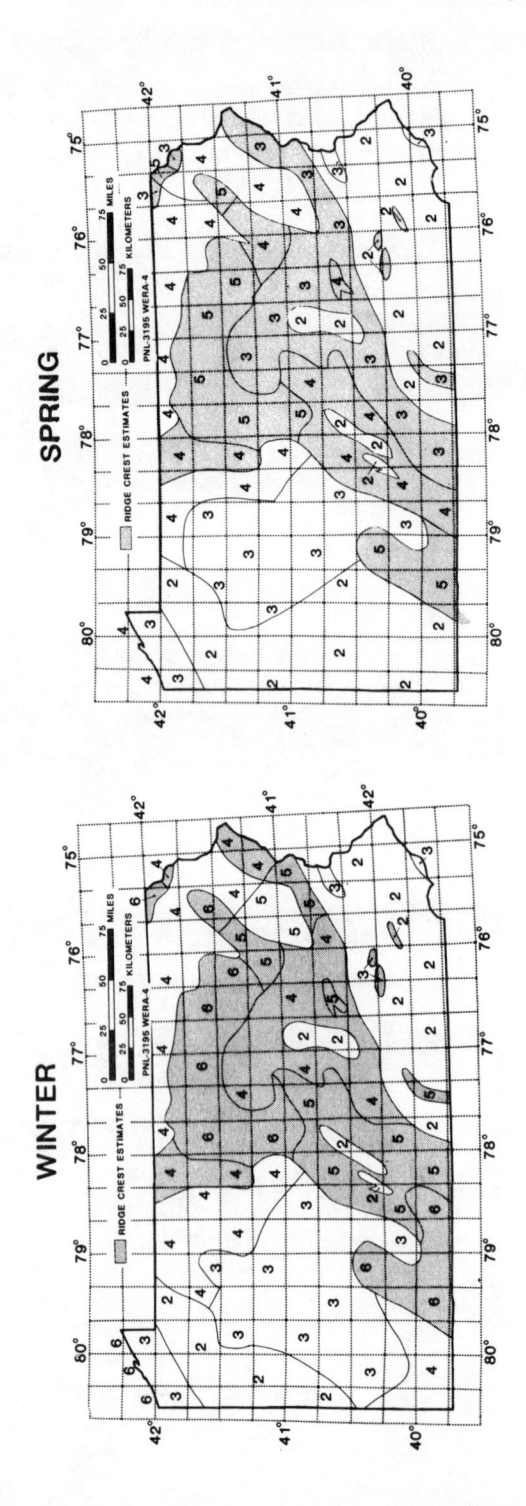

SPRING

Source: Battelle Memorial Institute, *Wind Energy Resource Atlas, Vol. 4—The Northeast Region,* September 1980.

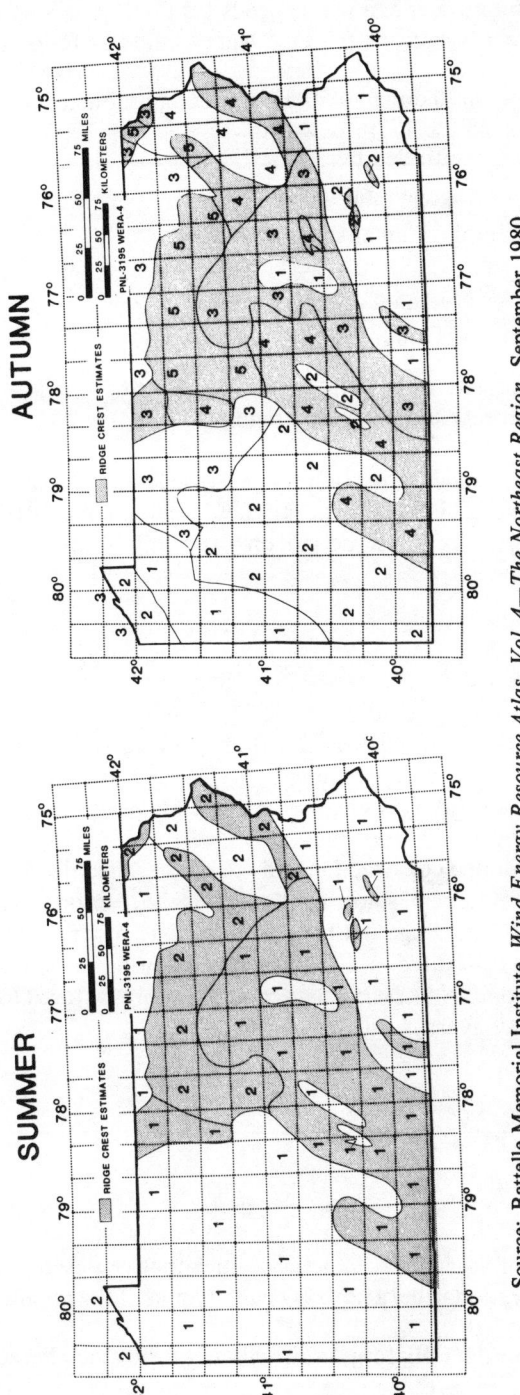

Figure 5–2. Seasonal Average Wind Power in Pennsylvania (Winter, summer, spring, autumn)

sites. Table 5-3 shows the monthly mean wind speeds for the Boone site at an elevation of 82 m. The hourly data on the tape were treated as follows:

1. Data given at 82 m was factored down to 50 m to be comparable to wind-atlas data at that level, using the 1/7th power law.
2. Data on the tape was then multiplied by:

$$\frac{\text{Class 4 average annual speed @ 50 m}}{\text{Boone average annual speed @ 50 m}}$$

or

$$\frac{\text{Class 6 average annual speed @ 50 m}}{\text{Boone average annual speed @ 50 m}}$$

3. Finally, the result was factored to the MOD-2 hub height of 65.6 m using the 1/7th power law. The result is shown on figure 5-3.

While classes 4 and 6 are defined in terms of a range, we selected the upper end of the range in each case. This is realistic in view of comments in the wind atlas that the maps show a lower bound on expected wind-energy potential.

The Wind Project

The wind project for simulation is based on the Boeing MOD-2 machine whose general characteristics are shown in table 5-4. The operating curves for the single machine and the 1000-MW cluster have already been shown in figure 3-3.

Table 5-5 shows the simulated construction and operation schedule. This is considered a realistic development since a utility is likely to proceed in stages, not proceeding to the next stage until the previous one proves itself. Figure 5-4 shows the energy production obtained by simulating the initial 100-MW array using the Boone, N.C., wind data, appropriately factored for the various classes defined by average wind speed.

Synthetic-Utility Characteristics

Synthetic-Utility Concept

EPRI has developed a set of so-called synthetic utilities that represent the wide array of utility configurations characterizing each region of the country [3].

The following, taken from the EPRI report, briefly describes the characteristics of each:

Table 5-3
Monthly and Annual Average Wind Speeds at Boone, North Carolina
(*height = 82m*)

Month	Wind Speed (m/s)
January	11.1
February	9.6
March	8.7
April	8.6
May	6.3
June	6.1
July	6.2
August	5.6
September	5.7
October	7.8
November	8.5
December	8.5
Annual average	7.73

Note: To estimate speed at 50 m, multiply figures in second column by 0.975.

System A. A summer-peaking system; largely coal and nuclear generation; predominantly a 345-kV (and lower) transmission network, being overlayed with 500kV and 765 kV; "medium-" length transmission lines. Loads are evenly distributed throughout the service territory.

System B. A winter-peaking system with high summer peaks; significant hydro generation; predominantly a 500-kV and 230-kV transmission network; long (average) transmission lines. This implies that generation is remotely situated from the load.

System C. A summer-peaking system; largely coal and nuclear generation; predominantly 345-kV and lower transmission; very long transmission lines. In this system, generation is very remote from the load.

System D. A summer-peaking system; predominantly oil generation, with some nuclear and coal; predominantly 500-kV and 230-kV transmission network; dense load distribution with generation situated near load. Hence average transmission-line lengths are short.

System E. A summer-peaking system; predominantly gas generation, with new coal and nuclear; predominantly 345-kV and lower transmission network with some 500 kV; uniform load distribution and medium transmission-line lengths.

System F. A summer-peaking system with high winter peaks; contains mostly oil and nuclear generation; 500-kV and lower transmission network; uniform load distribution; and medium transmission-line lengths.

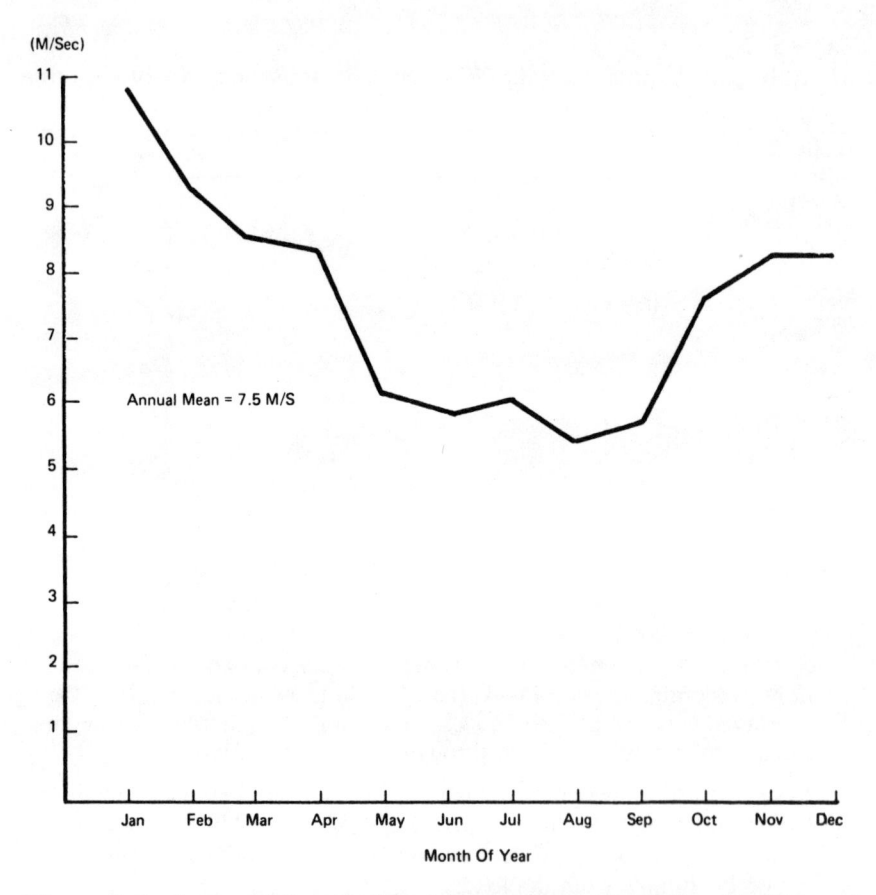

Figure 5-3. Average Monthly Wind Speed, Synthetic-Utility Site [Class 4 at MOD-2 Hub Height of 65.6 m (200 ft)]

The synthetic utilities are generic, that is, they capture the key features of "typical" utilities in each region of the country without applying exactly to any specific utility. This makes them extremely suitable for policy-system studies because they are neutral; utility executives and others intimately involved in utility planning and regulation felt they could comment frankly and impartially about various model results without appearing to take a position with respect to any particular utility. Furthermore, the EPRI description of the synthetic utilities is very detailed, allowing one to perform a realistic computer simulation of the electric-production process for a utility that is internally self-consistent in the sense that the utility's generation mix logically matches the electric load it must serve.

We have chosen synthetic utility D as an approximation of conditions in the industrial Northeast.

Table 5-4
Wind-Energy System for Simulation

MOD-2 wind turbine
2.5-MW output at 12.4 m/s
Hub height = 200 ft
Blade diameter = 300 ft
Minimum 10 diameter cluster spacing
Capital cost: approximately $800/Kw (1979 dollars)
O&M: approximately $8000/MW
Outage: 10 percent of time
Life: 30 years

Load

Figures 5-5 and 5-7 provide an overview of the monthly peak load and demand variations for synthetic utility D. These figures show that this utility peaks sharply in the summer due to air-conditioning demand. On a diurnal basis the load peaks in the late afternoon and early evening, which is typical of American utilities. We assume that load will increase at a compound annual growth rate of 2 percent, starting with 8244 MW in 1985.

Generation Mix and Expansion Plan

Table 5-6 shows the initial 1985 and the final 1999 equipment mix for synthetic utility D. Comparison with the 1979 utility mix of the northeast region shows that the synthetic utility is an excellent representation.

The EPRI expansion plan for this utility was based on an 8-percent compound annual load growth. However, figure 5-6 reflects an expansion plan for a 2-percent load growth using an approximation method for optimal

Table 5-5
Wind Cluster: Construction and Operation Schedule

Year	Construction (MW)	Operation (MW)	Capital Expenditures (Current $)
1984	100	—	$ 94.34 × 10^6
1985	—	100	—
1986	300	100	$318.00 × 10^6
1987	—	400	—
1988	300	400	$357.00 × 10^6
1989	—	700	—
1990	300	700	$401.5 × 10^6
1991	—	1000	—

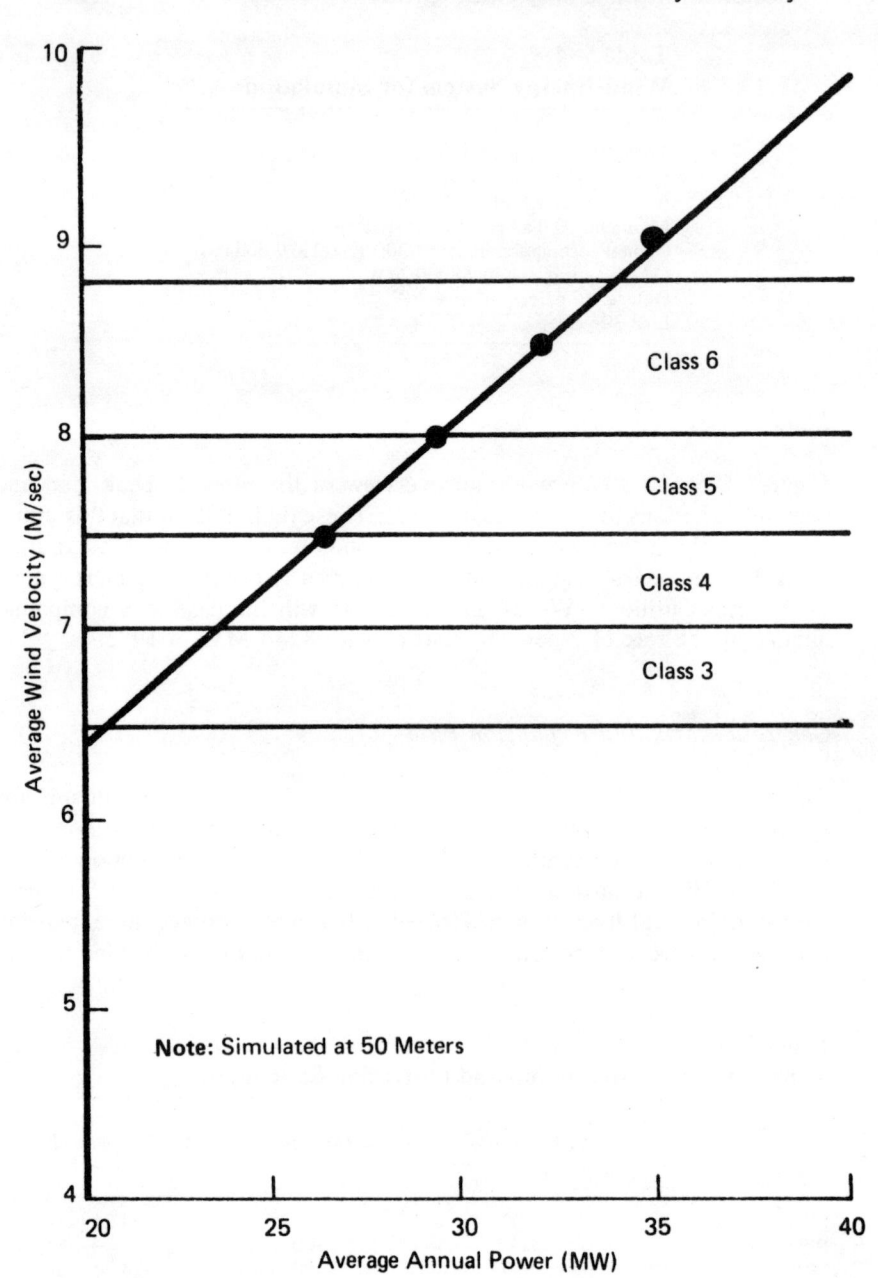

Source: Arthur D. Little, Inc.

Figure 5–4. Simulation Results Speed versus Power Output (100-MW array, Boone, North Carolina, distribution)

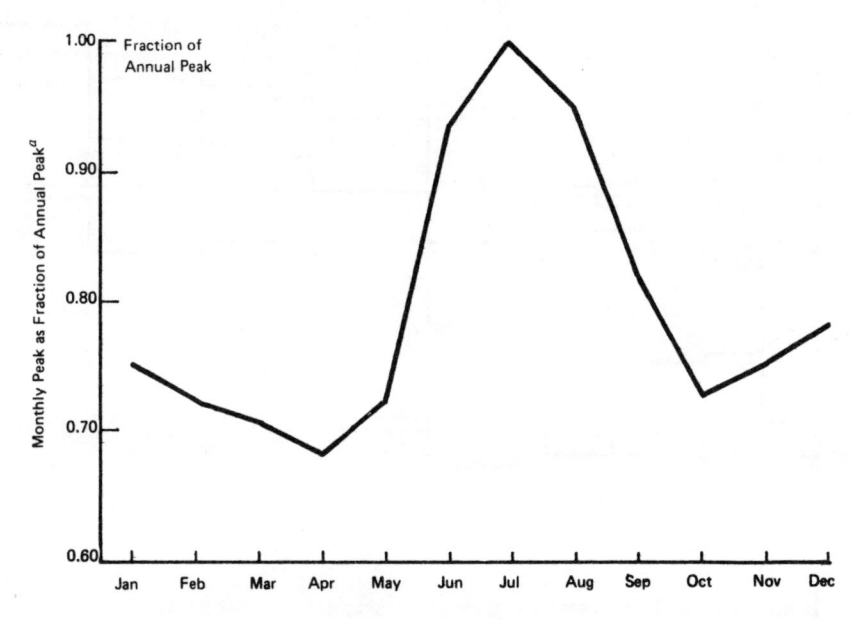

Note: This escalates at 2 percent compound rate through 1999.
[a] 1985 annual peak load = 8244 MW.

Figure 5-5. Monthly Peak-Load Variation

expansion planning. (See Winer [4].) We noted that nuclear plants would be the most economic units to introduce, but we assumed that only one such plant would come on line in 1986 and one other in 1998. This reflected our opinion that existing institutional barriers to nuclear expansion will continue to be felt until the 1990s, when planning for nuclear will begin again. The 1986 addition reflects a nuclear plant already in the pipeline. Thus, to meet load growth and replace retired capacity, the utility is seen to make an increasing commitment to coal capacity.

Fuel Prices

The fuel-price scenario used in the simulation is shown in table 5-7. There is an order-of-magnitude difference between displacement of the cheapest and most economic fuel. Clearly the increasing nuclear and coal production through 1999 will reduce the opportunity that the wind project has to displace the more expensive oil fuels.

Since the value of wind power is very sensitive to the fuel-price assumptions, we present a discussion of how the scenario fuel prices compare with recent prices reported in the northeast region.

Figure 5-6. Synthetic-Utility Expansion Plan

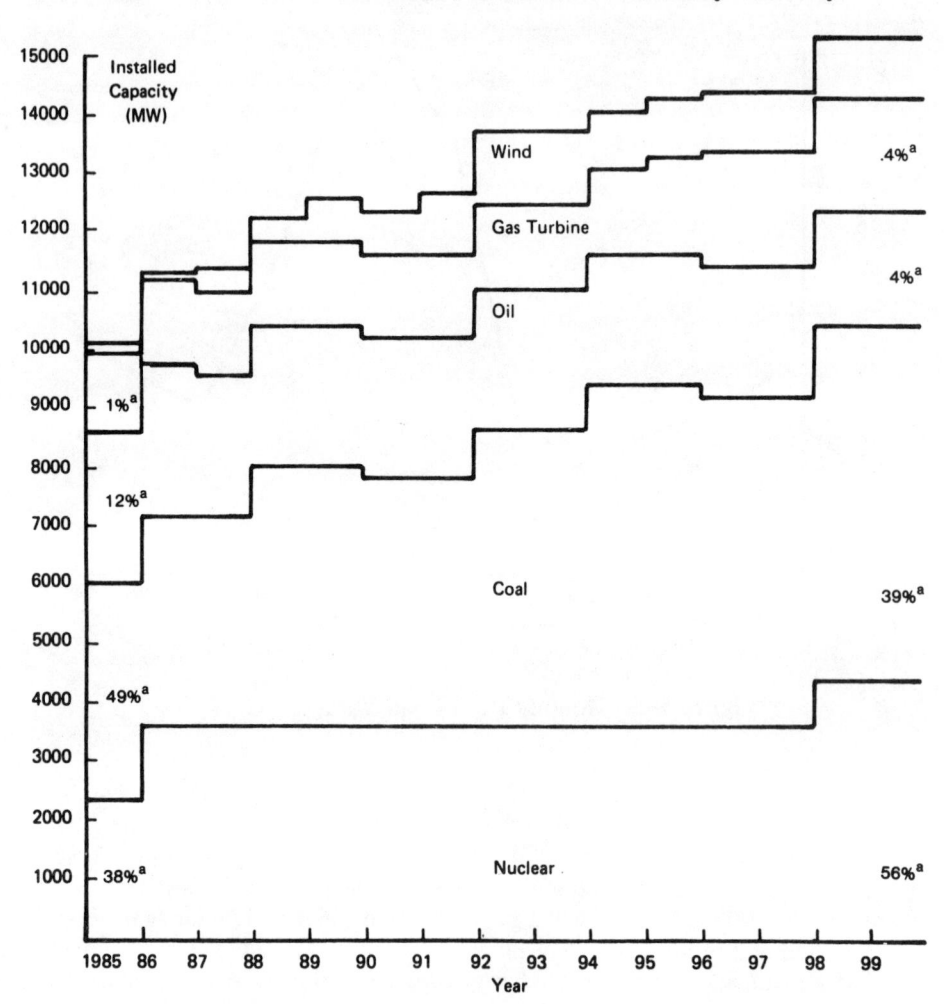

[a]Percent of total generated (without wind).

Table 5-8 shows how fuel prices have grown nationally since 1973. The estimates of table 5-7 are also shown on table 5-8 in ¢/10⁶Btu. The scenario residual-oil price is close to the national average for 1980. Our distillate-oil scenario price is less than the actual 1980 figure because of rapid price increases in this commodity that were not anticipated. The scenario coal price is more than the national average, reflecting eastern coal prices in the region. The national average includes low-Btu coals and lignite.

Table 5-9 focuses on fuel prices in the northeast region as of July 1980,

Constructing a Utility Case Study

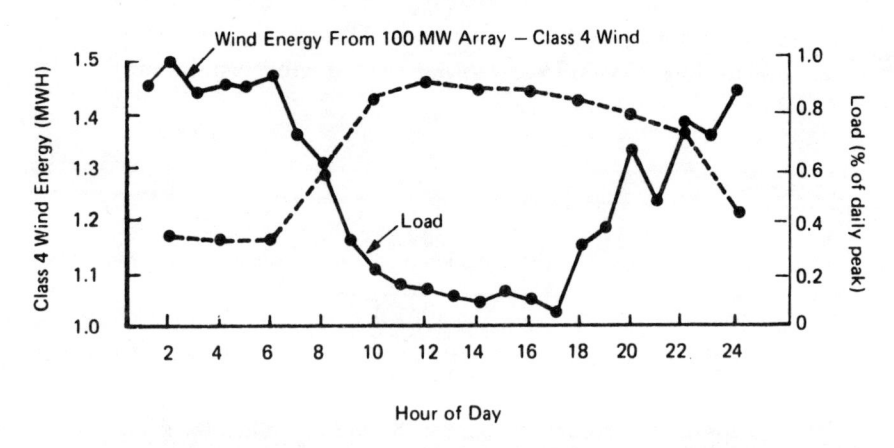

Figure 5-7. Diurnal Load versus Wind Velocity

the latest data available from DOE at the time we prepared input data for the simulation. The coal price in the scenario was set higher because current average prices reflect long-term contracts written over the last several years. The 220¢/10⁶ Btu scenario price is well below the 300¢ figure quoted on some new contracts. The scenario distillate-oil price is about 14 percent lower than we may have justified but is reasonable as a conservative basis.

Table 5-10 shows that coal prices remained steady between July and November 1980, as did distillate oil, but residual-oil prices rose sharply.

Table 5-6
Comparison of Synthetic-Utility Equipment Mix with Regional Equipment Mix in 1979

	Coal (%)	Nuclear (%)	Hydro (%)	Oil (%)	Gas (%)	Other (%)	Total (GW)
Region 1 New England	2	21	13	63	0	0	20.8
Region 2 N.Y./N.J./Pa.	9	13	11	67	0	0	48.0
Region 3 Pa./Del./Md./W.Va./Va.	53	14	4	28	0	0	68.9
U.S. Total	39	9	12	26	13	1	588
Synthetic Utility, 1985	36	24	0	40	0	0	10.1
1999	39	33	0	28	0	0	14.4

Source: Department of Energy, *Inventory of Power Plants in the United States,* April 1979.

Table 5-7
Generating-Fuel Costs, 1985 (Initial year of wind installation)

	Generation Costs (Mills/KWH)	Fuel Costs
Nuclear	13.5	$ 1.30/10^6 BTU
Coal	31.0	$86.10/Ton
Oil, residual	55.4	$33.48/Barrel
Oil, distillate	129.0	$40.07/Barrel

Note: Escalation to 1985 at 8 percent from 1980.

This rise continued through January 1981 but has fallen back slightly since that time.

Finally, it is instructive to look at spot-price reports in the region, as shown on table 5-11, to compare with the scenario assumptions. The scenario coal and residual prices are well within the reported range, while distillate is only about 8 percent lower than the 560¢/10^6 Btu quote that week.

Table 5-8
Cost of Fossil Fuels Delivered to Steam-Electric Plants
(¢ per million Btu)

	Coal	Residual Oil	Heating-Oil Wholesale
1973	40.5	78.8	—
1974	71.0	191.0	—
1975	81.4	201.4	—
1976	84.8	195.9	248.8
1977	94.7	220.4	282.9
1978	111.6	212.3	294.8
1979	122.4	299.7	443.0
1980 (Jan.–Sept.)	134.1	366.2	629.2
1980 } scenario	220	367	514 } 8% compound growth
1985 }	324	540	756 }
Rate of increase			
1973–1980	18.7%	24.5%	—
1975–1980	11.2%	11.5%	—
1979–1980	9.6%	22.2%	—

Source: Department of Energy, *Monthly Energy Review* (DOE/EIA-0035(81/01)), January 1980.

Note: These are national averages. Prices in the northeast region were generally higher, particularly for coal.

Table 5-9
Regional Fuel Prices, July 1980
(¢/10⁶ Btu)

	Coal[a]	Residual Oil	Distillate Oil
New England	160.0	366.3	618.9
Middle Atlantic	144.4	400.7	605.2
South Atlantic	156.1	362.7	594.5
1980 scenario	220	367	514

Source: Department of Energy, *Cost and Quality of Fuels for Electric Utility Plants,* July 1980. (Released for printing November 3, 1980.)

[a]Range of coal prices was 45–300¢/10⁶ Btu. These depend on the nature of the long-term contracts.

Table 5-10
Regional Fuel Prices, November 1980
(¢/10⁶ Btu)

	Coal	Residual Oil	Distillate Oil
New England			
(11/80)	160.5	489.1	628.6
(1/81)[a]	164	542	—
Middle Atlantic			
N.Y./N.J./Penn.	143.3	482.7	623.4
South Atlantic			
Del./Wash., D.C./			
Va./W.Va./Md./			
Fla./Ga./N.C./			
S.C.	158.2	475.1	597.4
1980 Scenario			
for Synthetic			
Utility D	220	367	514

Source: Department of Energy, *Cost and Quality of Fuels for Electric Utility Plants,* November 1980. (Released for printing March 1981.)

[a]Telephone report May 8, 1981, on DOE figures to be published.

Simulation of Utility-Operating Characteristics

Results of Simulation

Using General Electric's MPS computer program (as described in Chapter 3) we simulated the annual production costs of the synthetic utility for the following cases:

Table 5-11
Typical Fuel Prices Paid by Utilities in Northeast Region in April 1980

High-sulfur coal:
  (Pa. and R.I.)   90–208¢/10^6 Btu
Low-sulfur coal:
  (Del.)   192–230 ¢/10^6 Btu
#6 fuel oil: 269–561 ¢/10^6 Btu
#2 fuel oil: 560–615 ¢/10^6 Btu

Prices vary according to existence of long-term contracts.

Prices assumed for scenario building, 1980
Coal: 220¢/10^6 Btu ~ 26.6 × 10^6 Btu/ton → $54.26/ton
Residual Oil: 367¢/10^6 Btu ~ 6.2 × 10^6 Btu/bbl → $21.08/bbl
Distillate Oil: 514¢/10^6 Btu ~ 5.3 × 10^6 Btu/bbl → $25.25/bbl

Source: Data from *Electrical Week*, McGraw Hill, July 28, 1980 and August 4, 1980.

No wind project

A wind project on class 4 sites

A wind project on class 6 sites

These results, covering 1985–1999, are summarized on table 5–12. The column showing "net savings" is a true measure of the utility's avoided costs for the wind project. These results were used as production-cost inputs into the financial model.

Because there was only one calendar year of wind data from the Boone site, each of the years, 1985–1999, was simulated using the identical time series of wind. It would be more realistic to account for year-to-year variations in average wind characteristics.

Table 5–13 shows the unit savings, or avoided costs, in class 4 and class 6 on a $/MWh basis. It is noteworthy that there is very little difference in the value of a kWh generated for the two sites. This is because the additional wind energy at the class 6 site continues to displace fuel of equivalent value—the point of diminishing return not having been reached.

Table 5–14 shows the relative fuel displacement on a common MWh basis for selected years. We note that coal is the dominant fuel displaced. While oil is displaced in smaller quantities, reference to table 5–7 shows that it continues to make an important contribution to the value of wind energy. Nuclear-fuel displacement is clearly of minor value. The variation in the value of distillate displaced is a result of the circumstances encountered in the simulation from year to year, as equipment mix changed. We also note that the percent of load met by wind declines in the later years, as the utility's expansion plan (figure 5–7) continues to reduce the opportunity of wind to displace fossil fuel.

Constructing a Utility Case Study

Table 5-12
Production-Cost Simulation Results: Savings
($ million)

	No WECS		WECS Class 6		WECS Class 4		Net Savings		Installed WECS
Year	Fuel	O&M	Fuel	O&M	Fuel	O&M	Class 6	Class 4	(mw)
1985	1219	159	1205	158.1	1209	159	14.9	10	100
1986	1211	162	1198	161.1	1201	162	13.9	10	100
1987	1252	167	1204	165.1	1215	166	49.9	38	400
1988	1363	185	1314	182.6	1326	184	51.4	38	400
1989	1499	201	1413	196.8	1432	199	90.2	69	700
1990	1682	215	1584	211.3	1606	213	101.7	78	700
1991	1878	233	1725	227.3	1758	230	158.7	123	1000
1992	2053	257	(1892)	(250.7)	(1924)	(254)	(167.8)	(132)	1000
1993	2249	281	(2079)	(274.1)	(2114)	(278)	(177.0)	(138)	1000
1994	2470	307	(2291)	(299.5)	(2328)	(304)	(186.1)	(145)	1000
1995	2735	335	2547	327.8	2587	331	195.2	152	1000
1996	3074	361	(2873)	(353.8)	(2916)	(357)	(208.5)	(162)	1000
1997	3422	391	(3208)	(383.2)	(3253)	(387)	(221.8)	(173)	1000
1998	3523	401	(3296)	(393.0)	(3344)	(397)	(235.1)	(183)	1000
1999	3727	416	3486	408.6	3538	412	248.4	(193)	1000

Note: Parentheses denote best estimate by interpolation; otherwise, each figure taken from computer output.

Table 5-13
Net Savings per Wind-Generated MWH
(current dollars)

Year	Class 4	Class 6
1985	49.000	49.372
1987	45.107	44.442
1989	47.302	47.608
1991	59.516	59.847
1995	73.486	73.521

Table 5-15 shows the equivalent fuel displacement in units of barrels for oil and tons for coal.

Observations

These results provide a basis for evaluating the engineering-economics of the project and inputs to a financial analysis of the investment.

We have attempted to make this base case as realistic as possible so that utilities with access to the requisite wind resources could relate the results to their own situation.

The base case required a combination of "liberal" and "conservative" assumptions that we briefly describe. Liberal assumptions are those that may be viewed as optimistic. These include:

The installed cost of the wind systems

The operating and maintenance costs of the wind systems

The choice of dispatch rule relative to spinning reserve. (See chapter 3.)

Conservative assumptions are those that may be viewed as pessimistic. These include:

The limitations imposed by the Boone, North Carolina, site in which wind availability is poorly correlated with the utility load both diurnally

Table 5-14
Fuel Displaced by Wind (10^3 MWH)
(class 4 wind case)

Year	Wind	Distillate	Oil	Coal	Nuclear	Percentage of Utility Load Met by Wind
1985	231	25	71	134	1	0.54
1987	918	22	210	585	101	2.06
1989	1574	11	259	1081	222	3.40
1991	2214	67	409	1490	246	4.59
1995	2214	32	299	1612	270	4.24

Constructing a Utility Case Study

Table 5-15
Fuel Displaced by Wind (Fuel Units)
(class 4 wind case)

Year	Distillate (Thousand Barrels)	Residual (Thousand Barrels)	Coal (Thousand Tons)	Wind (MW)
1985	74	112	43	100
1987	78	345	189	400
1989	43	449	346	700
1991	222	682	475	1000
1995	79	512	519	1000

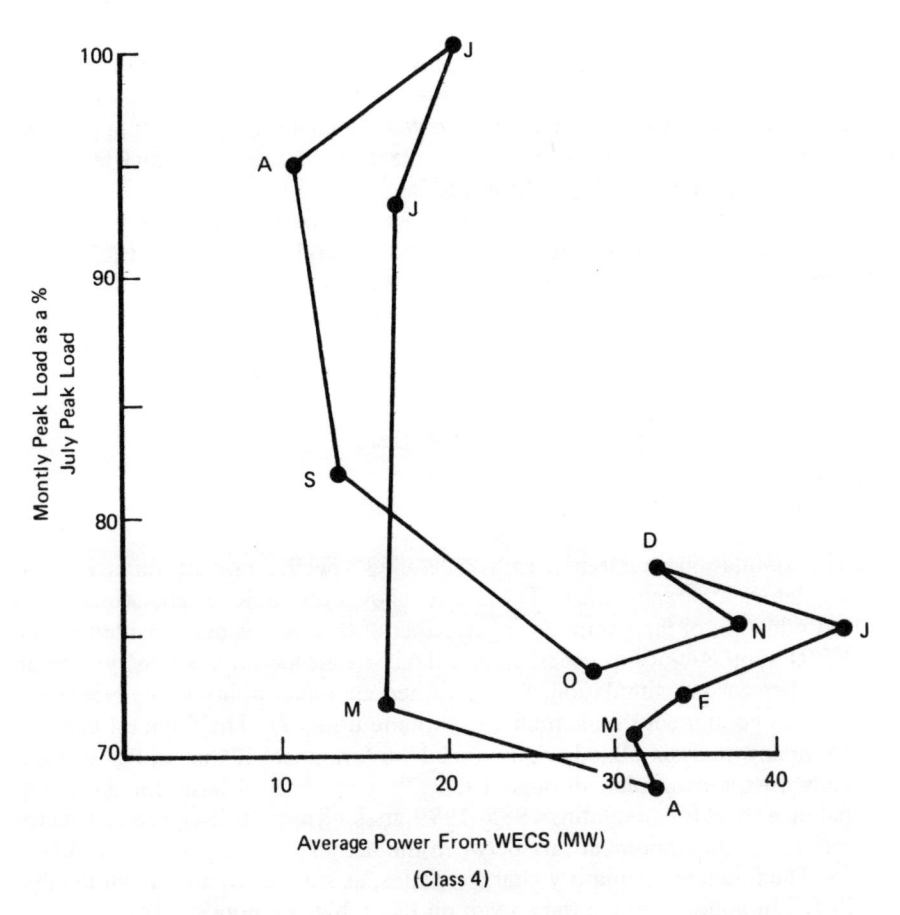

Figure 5-8. Monthly Variation of Wind-Power Output versus Peak Load

and seasonally (see figures 5–7 and 5–8). We believe that wind prospecting in the region will produce better sites.

The knowledge that it is possible to identify sites exceeding class 4 wind power within the area identified as class 4 or lower in the *Battelle Wind Atlas*.

The assumption that the utility's generation plan is independent of the wind project. This not only assures the minimization of fuel savings credited to the wind project but also precludes any capacity credit related to reduced capital expenditures.

The utility's heavy use of coal, which is more characteristic of utilities in the southern portion of the region. Use of a synthetic utility starting with a larger oil base would have greatly improved the value of wind.

A key element in the economic and financial analysis is the projection of future fuel unit costs and the rate of inflation. We used 6 percent for inflation of costs in general and 8 percent for fuel. It is difficult to characterize this as liberal or conservative over the fifteen year simulation horizon. It appears to be a fair and reasonable basis for project evaluation and consistent with general practice by utility planners today.

On the whole, we have tried to provide a reasonable balance between liberal and conservative assumptions. The liberal ones are prerequisite for wind energy to become a reality. The conservative ones represent the typical planners tendency to understate the case when there is uncertainty and lack of data.

Financial and Regulatory Characteristics

Basic Computer Input

The simulation requires a rather detailed specification of financial and regulatory characteristics. These are shown in table 5–16, which is a reproduction of the computer input. Most of this data is not provided by the EPRI synthetic-utility specification. Thus we needed a basis to synthesize this data for our simulation. We identified an actual utility in the Northeast with an equipment mix similar to synthetic utility D. The General Electric Company had simulated the financial performance of that utility using a known expansion plan through 1984. This provided a basis for the initial balance sheet for simulating 1985–1999, making appropriate adjustments for differences in equipment mix between the actual and the synthetic utility.

The financial regulatory characteristics, as simulated, are shown in table 5–16. In addition to the data given on this table, we note:

Constructing a Utility Case Study

A future general inflation rate of 6 percent was used in simulation.

A fuel-escalation rate for all fuels of 8 percent was used.

While table 5-16 shows a minimum rate of return on common equity (ROE) of 17 percent, this applied only to 1985; in most years we used 13 percent. This is explained later.

PU = public utility.

ITC = investment-tax credit.

TD = term debt.

LT = longterm.

CWIP = construction work in progress.

P.U. fuel rider = 1.0, means that a fuel-adjustment clause passes 100 percent of fuel-price increases onto the customer rate.

AFDC = allowance for funds during construction (that is, AFUDC—see p. 71).

LAG means that the customer rate is computed in the current year to meet the target return on equity but is not applied until the following year.

Regulation—common equity—means that the utility is regulated to achieve a return on common stock rather than on equipment in the rate base.

Setting Rate of Return Each Year

Typically, when a utility operates in a situation with regulatory lag, there is a potential financial strain during any year in which a large generation plant is brought on line. In that year the AFDC allowance drops to zero, but the plant is not yet factored into the rate base. We originally ran FSP with a target ROE of 13 percent in all years. During the years in which large coal or nuclear plants came on line (1986, 1992, 1994, 1998), we found that the earnings squeeze was very severe. During 1986 and 1998, the years in which the two new nuclear plants were introduced, the pretax bond-coverage ratio reported as output dropped significantly below 2.0, which is below the ratio usually required by utility-bond indenture provisions. We were forced therefore to provide our utility with the equivalent of "emergency rate relief" during these four years in order to avoid creating a hypothetical default situation.

"Emergency rate relief" was factored in by increasing the target ROE during the years prior to these four problem years. Because the regulation algorithm assumes rates in year N are based on conditions in year $N-1$, we

Table 5-16
Computer Input to the Financial Simulation Program (Sample)
(Input Data—$ in Thousands)

INITIAL BALANCE SHEET		
ELECTRIC PRODUCTION	3274568.	
TOTAL PLANT IN SERVICE	6005366.	
GAS PLANT IN SERVICE	0.	
CWIP, GENERATION PLANT	3019750.	
CWIP, OTHER PLANT	162956.	
CWIP, GAS PLANT	0.	
DEPREC. RESERVE, ELEC.	1880950.	
DEPREC. RESERVE, GAS	0.	
NUCLEAR FUEL INVENTORY	207861.	
CASH BALANCE, END OF YR	104703.	
ACCOUNTS RECEIVABLE	1154799.	
FOSSIL FUEL INVENTORY	416221.	
TOTAL INVENTORY EXCL. NUC.	518563.	
COMMON STOCK OUTSTANDING	1412569.	
RETAINED EARNINGS	1502324.	
PREFERRED STOCK	875677.	
LONG TERM DEBT + CUR. MAT.	3606005.	
SHORT TERM DEBT	157933.	
ACCOUNTS PAYABLE + MISC.	1069177.	
DEF. INV. TAX CREDIT, ACCUM.	313096.	
DEF. FED. INC. TAX, ACCUM.	356267.	

FINANCIAL DATA	
HISTORICAL INFLATION, OTHER	1.040
HISTORICAL INFLATION GAS	1.040
FINANCE WITH CASH ON HAND	YES
MINIMUM CASH POSITION	1000000.
MAXIMUM DEBT RATIO, PU	0.520
DIVIDEND PAYOUT RATIO, PU	0.650
COMMON STK OUTSTANDING	66569030
COMMON STOCK P/E RATIO	7.000
MAX. PREF. STOCK, PU	0.120
MIN. BOND ISSUE SIZE	100000.
MAX. BOND ISSUE SIZE	300000.
MIN. PREF. ISSUE SIZE	10000.
MAX. PREF. ISSUE SIZE	200000.
MIN. COMMON ISSUE SIZE	10000.
MAX. COMMON ISSUE SIZE	300000.
INCREASE ISSUE SIZES, AUTO	YES
FACTOR, SHORT TD/CAPITALIZ.	0.050
INTEREST FACTOR, SHORT TD	1.100
INTEREST FACTOR, PF DIV.	1.050
AFDC RATE/L.T. RATE	0.900
LONG TERM INTEREST RATE	0.090

INCOME STATEMENT
****

ELECTRIC REVENUE	2900000.
GAS REVENUE	0.
FUEL COST	1185294.
NET OTHER NON-OPER. INCOME	0.
OTHER PLANT OPER. EXPENSE, PU	0.075
GAS PLANT OPER. EXPENSE, PU	0.075

RETAINED EARNINGS
****

COMMON STOCK DIVIDENDS	209971.
PREF. STOCK DIVIDENDS	61415.

SYSTEM DATA
****

ENERGY GENERATED, GWH	41609.
LOSSES, PU OF ENERGY GENR.	0.070
HISTORICAL PEAK LOAD GROWTH	0.030
FIRST YEAR PEAK LOAD GROWTH	0.020

PLANT DATA
****

BOOK LIFE, OTHER PLANT	30
BOOK LIFE, GAS PLANT	30
IN SERVICE CURRENT YR, OTHER	0.25
IN SERVICE CURRENT YR, GAS	0.25
CWIP ELIG. FOR AFDC, OTHER	0.75
CWIP ELIG. FOR AFDC, GAS	0.75

REGULATORY DATA
****

P.U. FUEL RIDER	1.000
REGULATION	COMM. EQUITY
AMT. CWIP INCL. IN RATE BASE	0.
MINIMUM RATE OF RETURN, PU	0.170
MAXIMUM RATE OF RETURN, PU	0.170
DESIRED RATE OF RETURN, PU	0.170
REGULATION LAG	LAG

TAX DATA
****

FEDERAL INCOME TAX RATE, PU	0.460
COMPOSITE MUNICIPAL TAX, PU	0.050
REVENUE TAX RATE, PU	0.050
TAX DEPREC. LIFE, OTHER	30.
TAX DEPREC. LIFE, GAS	30.
INV. TAX CRED. (ITC), PU	0.100
MAX. ALLOW. CREDIT, (TAX LIAB)	0.
ITC ELIGIBILITY, OTHER, PU	1.000
ITC ELIGIBILITY, GAS, PU	1.000
ITC - PROGRESS PAYMENTS	YES
TAX SAVINGS METHOD	NORMALIZED
ITC AMORTIZATION LIFE	30

raised the target ROE during the years 1985, 1991, 1993, and 1997 to 17, 14, 14, and 17 percent, respectively. This forced the program to increase the customer rates somewhat during the four problem years, but even so, there were still significant drops in ROE during these years. In order to prevent corresponding drops in the dividends paid to common-stock holders in these years, the payout ratio was raised from 0.65 to 1.0, 0.75, 0.75, and 1.0 in 1986, 1992, 1994, and 1998, respectively. These changes became a permanent part of the input; in all the cases we discuss, the same rate relief was provided in these four years. Furthermore, in those three years in which 300 MW of WECS were brought on line (1987, 1989, and 1991), a similar (though less severe) drop in earnings occurred, but we did *not* provide rate relief during these years by raising the target ROE because we wanted a uniform baseline against which to examine the financial impacts of the wind project.

Conventional Engineering–Economic Analysis

In chapter 3, we discussed the literature in which breakeven cost analysis was used to evaluate an investment into wind energy. Table 3–1 displays some of the results of earlier investigations with our own. While it is not our purpose to prove the feasibility of a wind project for the synthetic utility, it is useful to examine the range of outcomes that are possible, depending on the assumptions used in the analysis. We thus present a simplified analysis of the breakeven cost based on the initial 100-MW wind project installed in 1985. For discussion purposes the following assumptions are made:

> The "go–no-go" criterion for breakeven cost is $1000/kW, in 1985 dollars. This assumes that the utility can install the wind machines at this price, including land and transmission costs.
>
> While unit fuel costs escalate at 8 percent per annum, the project will have to be justified on the basis of effective fuel-cost escalation, using simulation in which the utility's expansion plan into coal combines to significantly decrease the opportunity of wind to displace the higher-price fuels.
>
> The cost of capital is 9 percent, the federal tax rate is 46 percent, and state and local taxes are 5 percent of assets.

Computation of Breakeven Cost

The methodology for conventional engineering-economic analysis has been described in chapter 3 in the section, "Life-Cycle Model." Figure 5–9 is a

Constructing a Utility Case Study 119

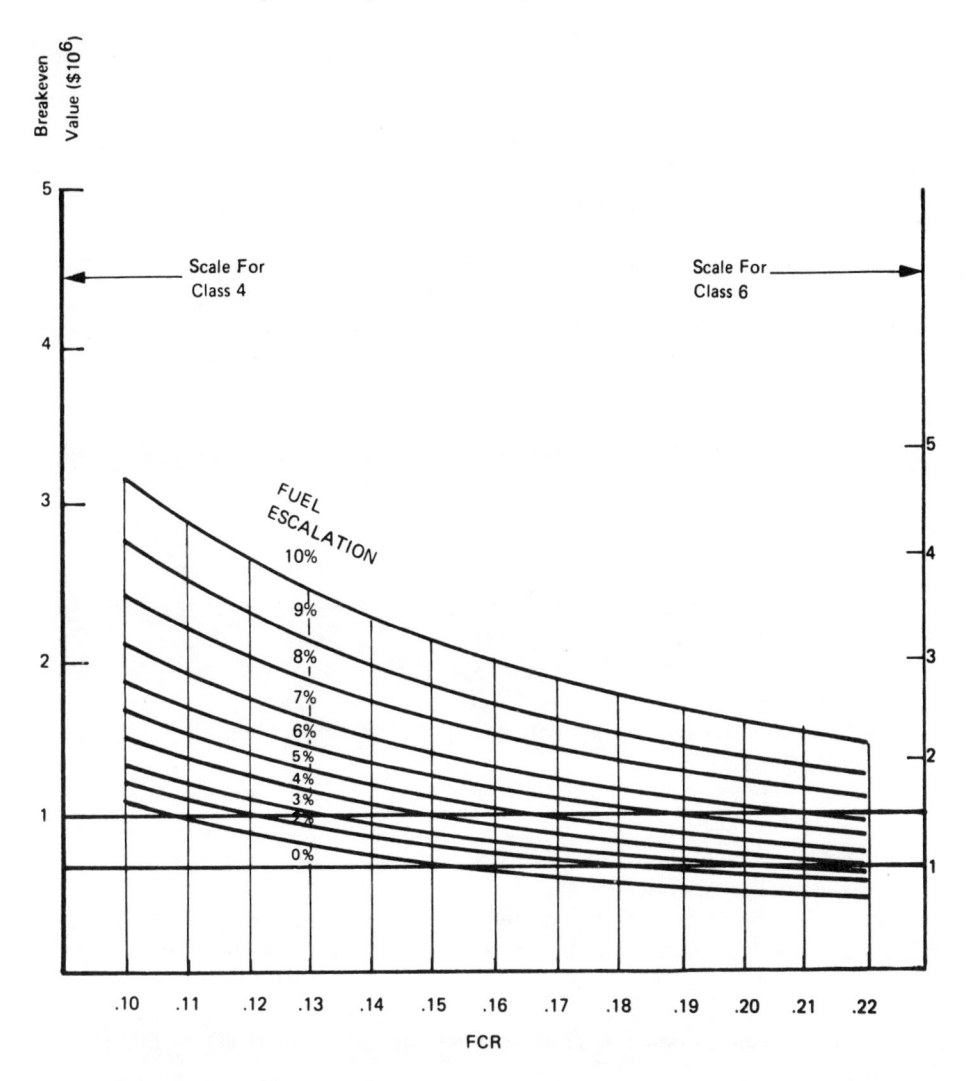

Figure 5-9. Breakeven Cost Results (Discount rate is 10 percent.)

plot of equation 3.1, in which the discount rate was set at 10 percent. The expression for the present value of a series of uniformly escalating annual value of fuel savings is given by:

$$PV = \left[\frac{1+g}{k-g}\right]\left[1\left(\frac{1+g}{1-k}\right)^N\right]S_0 \quad \begin{array}{l} g \neq k \\ g = k \end{array} \quad (5.1)$$

where

g = fuel escalation rate

k = utility-discount rate (10 percent)

S_0 = initial-year fuel savings

In figure 5–9 there are two scales for the breakeven value. The left scale is for the class 4 wind case and the right scale, class 6. The value of fuel savings from 100-MW installation in 1985 was determined by simulation to be $100,000 for class 4 and $149,000 for class 6. Thus for any combination of fuel escalation rate and FCR, figure 5–9 displays the breakeven cost in a class 4 or a class 6 situation.

While the unit price of fuel escalated uniformly by 8 percent in the scenario, the annual dollar value of total fuel savings per installed megawatt was seen, when simulated to escalate in a nonuniform manner, and at an effective rate of less than 8 percent (see table 5–12). This is because the simulation increased the installed wind from 100 MW to 1000 MW over the first eight years, and the changing utility equipment mix was reducing the opportunity of wind-generated power to displace the higher-cost fuels. Taking the 1985 and the 1999 values only, the approximate escalation rate of savings for class 4 case is only 4.8 percent, and for class 6, 3.7 percent, for the first fifteen years, respectively. [FCR = 0.20 when computed at $\tau = 46$ percent, $\beta = 0.05$.]

Entering figure 5–9 with FCR = 0.20, and fuel escalation at 4.8 percent, the class 4 breakeven cost is less than 10^6, the minimum value to justify the project. Shifting to the scale for class 6, however, 3.7 percent fuel escalation gives a small excess over 10^6.

It has been assumed that the effective fuel escalation for class 4 case remains at 4.8 percent for the entire thirty-year project life. This may be overly conservative. If escalation is at 4.8 percent for the first fifteen years, and 8 percent for the next fifteen years, the effective uniform rate over thirty years becomes 6.4 percent. This yields a project that just meets the minimum breakeven value in figure 5–9.

Factors That Increase Estimated Breakeven Cost

The effective growth rate in savings would be increased by any one of the following:

The 100 MW in 1985 was the only wind cluster introduced. (The effective rate would have been close to 8 percent.)

The unit fuel costs escalated at more than 8 percent per annum.

The initial-year fuel cost was higher (linear increase in breakeven cost).

The utility failed to achieve the expansion plan that shifted fuel consumption away from oil.

The simulation used data available through September 1980 to estimate fuel prices. For example, 367¢/10^6 Btu for residual-fuel oil appeared a reasonable estimate at the end of 1980 when the simulation was set up for the computer (table 5–8). By November 1980, however, residual oil in the region was about 480¢/10^6 Btu (table 5–10) and by January 1981 reached 542¢/10^6 Btu, an increase over our estimate of 30 percent. The effect on the breakeven value for Class 4 would be to significantly exceed the criterion value of 10^6/mw.

The simulation assumed an 8 percent unit fuel-cost escalation resulting in a 4.8 percent effective escalation in savings. Table 5–8 shows that unit fuel costs for residual oil actually escalated at 11.5 percent for six years *after* the price had already more than doubled from 1973–1974. Assume that unit fuel costs in simulation escalated at 10 percent and yielded 7-percent effective escalation in savings. Figure 5–9 now shows the project meeting the breakeven cost criterion at FCR = 0.20.

Clearly, the results of engineering-economic analysis are very sensitive to basic assumptions about initial-year fuel price and escalation rate. In a period of inflation and unstable fuel prices, it is very difficult to defend these assumptions. In the simulation, we chose a combination of reasonable assumptions that yield a realistic perception of the breakeven value of the project.

Increasing Breakeven Costs by Policy Changes

As reported in chapter 3, previous studies have developed the quantitative impact of policy changes on the breakeven cost (or, analogously, on the levelized busbar cost). Any policy that reduces FCR in figure 5–9 clearly increases breakeven cost. These may include:

Reducing effective federal income-tax rate on project

Reducing the cost of borrowing

Reducing local taxes

Reducing tax life (accelerated depreciation)

Reducing the discount rate also increases the present value of the fuel saved. Incentives such as a direct subsidy or an increased investment-tax credit have the effect of decreasing the criterion value (installed cost) with which the breakeven value is compared.

Thus, it may be argued that if the project breakeven value fails to exceed the installed cost, one or more of these incentives may be invoked to improve the economics of the project. Literature cited in chapter 3 took this point of view.

Project Decision Factors Using Breakeven Cost Criteria

For the class 4 case, with no incentives to reduce FCR or the installed cost and for the fuel prices used in the simulation, our synthetic utility would find that the full project at 1000 MW is "no-go" based on the single criterion that breakeven cost must exceed installed cost. However, a project limited to the initial 100 MW may well meet the breakeven criteria. For the class 6 case, the full project would be a "go" based on the same criterion.

Given these results, the synthetic-utility action is likely to include:

Examining the wind resource more closely to identify sites that resemble class 6 in wind speed or that have wind histories better correlated with load than the Boone, North Carolina, tape indicated.

If possible, develop the first 100 MW on sites that yield fuel savings approximating the class 6 case.

Assuming that initially 100 MW could be justified and that all subsequent development will have to be on class 4 sites, a decision to "go" will be contigent upon:

Continued surges in the cost of fuels between now and the time to commit to the project. (We have already seen that the fuel-price increase that occurred in the several months following the computer run would have had this effect.)

Federal or state actions that reduce FCR or installed cost.

Thus, a decision for a significant wind program needs to be supported by a strategic planning process in which each increment of investment is justified by the latest available data, within a comprehensive evaluation framework.

In addition to meeting the breakeven cost criterion (under a set of reasonably conservative assumptions about fuel savings, installed costs, and so on), there are other criteria that are important to utilities. These include:

Confidence that the equipment will indeed perform for thirty years

Confidence that O&M costs will not become unmanageable (as when blades or other components fail)

Constructing a Utility Case Study

Confidence that the machines will meet targeted capacity factors (no excess downtime)

Confidence that the intermittent electric energy can be safely integrated with the grid

Ability of manufacturer to deliver the equipment and stand behind it

Community acceptance of the physical presence of wind turbines on the landscape

Acceptance of a given level of environmental impact, with particular reference to bird collisions, electromagnetic interference, and noise propagation

The utility will also weigh the following considerations into a "go" decision, even if breakeven cost is less than the installed cost, particularly for a demonstration up to 100 MW.

A desire to assert a leadership role with respect to wind-power development.

The positive public image associated with a high-visibility renewable-energy program.

The ability to accept the risks associated with the initial 100-MW development. Should it fail to meet expectations, the losses would be limited.

A desire to gain operating experience with a technology that may play an increasing future role in electrical-energy generation.

Pressure from its regulatory agency.

Let us assume that all these criteria can be satisfied and that over the several-year period in which plans are made, the utility will identify the required number of sites to ultimately install 1000 MW of wind power. Before committing to a program of this magnitude, the utility will want to examine the impact of the investment on the financial condition of the utility itself. This requires tools of analysis that go beyond engineering-economics. This is developed in the following chapter.

References

1. Battelle Pacific Northwest Laboratory. *Wind Energy Resource Atlas: Volume 4, The Northeast Region,* PNL-3195 WERA-4/UC-60. GEOMET Technologies Inc., Contractor, September 1980.

2. Battelle Pacific Northwest Laboratory, Atmosphere Sciences Department. *Meteorological Data Tape for Boone, N.C.* October 1978–December 1979.
3. Electric Power Research Institute (EPRI). *Synthetic Electric Utility Systems for Evaluating Advanced Technologies.* Power Technology Inc., Contractor, February, 1977.
4. Winer, Bette M. Arthur D. Little, Inc. *Testimony of Dr. Winer before the Federal Energy Regulatory Commission.* Project No. 2424, May 4, 1981.

6 Financial and Policy Analysis of Scenarios

This chapter begins with a background discussion that describes the context for interpreting the financial-model results. The sensitivity to certain key assumptions about the utility and its financial and regulatory environment is indicated and some of the methodological limitations are discussed.

The background section is followed by a policy-oriented discussion of the specific scenarios analyzed. It includes a description of each scenario, the policies it embodies, and the rationale for including it. An analysis of the results of each scenario simulation is presented in terms of (1) the sensitivity of key utility financial parameters to certain policy changes, and (2) the likely effects of policy changes on a utility's decision to invest in wind turbines. The major public-policy issues raised by the scenarios are also described.

Specific recommendations for policy changes are not made, as these require broader issue analysis beyond the scope of this study, which concentrates on the usefulness of systems-analysis methodology. However, we do note the policy options that appear to have the most salutary effects on the utility. While the methodology is powerful, it reflects only some of the criteria for a complete policy evaluation.

Issues in the Case Study

The results of an investment into wind energy are inextricably linked to the utility's financial condition, expansion plan, and generating mix, as well as to the financial, economic, and regulatory environment in which it operates. This section discusses these issues and the sensitivity of our results to them.

One reason for having designed a base case that assumes a relatively large amount of investment into wind projects is to achieve a level of fuel savings and investment-financing requirements large enough to impact the utility's financial condition. In this way we are able to study the effect of policy variations on the utility's cash flow, balance sheet, and financial planning. It should also be noted, however, that even with 1000 MW of wind, some of the effects of the WECS investment are diluted by the size of the overall system and have to be inferred.

The base case reflects the current financial and regulatory environment for utilities: one that is not conducive to utility investment in general, let alone investment in renewable-resource technologies. Summarizing from chapter 4, the following factors have contributed to the poor investment climate for utilities:

High rates of inflation coupled with regulatory lag and high fuel costs

Poor capital markets in general and for utilities in particular

Very low price-earnings ratios for utility stock and corresponding low market or book-value ratios

Public attitudes that are generally hostile to utilities

Rapidly changing federal regulations (for example, environmental requirements, PURPA, Fuel Use Act)

The combination of these factors results in a base case that is not really a desirable norm for investment in wind technology. While this is realistic, there may be some long-term distortion inherent in the evaluation of policy scenarios that build on an adverse base case. In fact, it appears that in most instances, the policy incentives related to the wind investment are insufficient to overcome the negative bias of the base case. The superimposing of a large additional construction requirement on a utility under the assumption of a poor investment climate inevitably penalizes the desirability of such an investment.

There are other assumptions in our analysis that bias the financial results against investment in wind turbines, a major one being the conservative assumptions about future fuel costs compared to what utilities have recently experienced. Also we have not allowed a capacity credit for wind turbines, reflecting a current utility-industry viewpoint, although in the long term a capacity credit will probably become accepted. While this assumption is a reasonable one in light of the apparent risk that utilities perceive in investing in wind equipment, it does result in possible overstatement of new financing requirements. If a capacity credit were granted, the utility would require less capital to support the expansion plan. A lower investment requirement would result in lower financing costs, less common stock issued, improved price-earnings ratios, lower revenue requirements, and lower customer rates. It is only in granting wind equipment a capacity credit that the full benefits of wind investment can ultimately be realized.

Some of our assumptions may be considered optimistic. We have assumed that wind-generating equipment can be brought on line in a relatively short time compared to conventional baseload equipment, the lead time assumed to allowance in the rate base being two years. If, however, there are regulatory approval issues to be resolved, based on licensing (for example, environmental or safety) issues raised by intervenors, the resulting delays could be devastating to wind-turbine investment just as they have been in the case of other generating capacity.

We assumed that the utility would be able to site wind turbines and find land for 1000 MW. Whether this assumption is warranted can be determined only be research on a utility-by-utility basis. For example, wind conditions and site availability in Hawaii, where a major investment in wind turbines is taking place, is dramatically different from New England with its denser population and multiplicity of jurisdictions. The effects of problems in siting and land availability would be to raise the capital cost per kilowatt. Moreover, real-estate speculation in sites may result in rising land prices for areas with good wind resources. There are already reports of speculation of this type in California and the Middle West.

There are, of course, inherent limitations in the precision of this or any similar model. As a result, there is some distortion in comparing a scenario involving wind against a base case that has no wind. (Comparisons of the various scenarios involving wind among themselves involve no such distortion.)

The base case reflects a demand growth of approximately 2 percent per year. This assumption has the following effects:

> To the extent that the assumed demand growth proves too high because of present conservation trends and elasticity responses to high prices, the utility financial position would be better than projected because its investment requirements would be lower.

> If demand growth proves higher than 2 percent, the utility will be worse off than we have estimated because it will incur additional financing requirements.

The expansion plan that we have defined reinforces the effects of the demand-growth assumptions. It results in the displacement (by wind) of lower-cost fuel at the margin (coal and nuclear). If demand growth is lower than 2 percent, not all of this expansion would be needed, and proportionately more oil would be displaced than coal or nuclear fuel. The additional displacement of oil along with the lower investment requirement would improve the utility's financial position.

Our estimates of the capital cost of wind investments reflect the optimistic assumption that commercialization of wind-generating facilities will take place in the near future, thereby reducing the cost per kilowatt in real terms as a result of economies of large-scale production. This may require that government continue to play a role in the research, demonstration, and commercialization processes. Given the recent changes in Washington in the administration of renewable-energy policy, it is difficult to assess the federal role in the future.

Thus we have tried to highlight some of the factors related to uncertainty and limitations of the model. This underscores the need for experience and good judgment in interpreting the outcome of this or any related systems-analysis undertaking.

Selection of Policy Alternatives for Study

Definition of Base Cases

Later in this chapter we present table 6–2, which contains a summary of the financial-model results for eighteen separate scenario runs. These begin with a base-case situation in which the utility is regulated on common equity, there is a one-year regulatory lag, and there is a 100-percent fuel-adjustment clause (scenario 1). Then a wind investment of 1000 MW is imposed with no policy changes (scenario 2), and eight policy variations to this base case are presented (scenarios 3–10).

To study the effect of zero regulatory lag, we introduce a new base case without wind (scenario 11), and add wind at two levels of wind energy (scenarios 12 and 13). Scenario 13, termed *supercase*, is one with extremely optimistic fuel-cost savings to illustrate how the regulatory process allocates the savings.

To study the effect of putting construction work in progress (CWIP) into the rate base, we create a third base case without wind in which the utility is regulated on rate base rather than on common equity (scenario 14). We then add wind (scenario 15) and put CWIP into the rate base (scenario 16).

Finally, we create a fourth base case with no wind in which the fuel-adjustment clause is eliminated (scenario 17). We then impose the wind investment into this situation (scenario 18).

Definition of Policy Scenarios

In deciding on these eighteen runs of the financial model, we went through a process of analyzing policies and decided how to group them and represent them as computer input. We began by categorizing according to the dominant effect of the policy:

1. Higher return on equity
2. Improvement of cash flow
3. Government subsidy
4. Institutional/regulatory changes

Clearly, there is overlap among these categories, but they offer a convenient frame of reference for the discussion that follows.

Several policy incentives may have similar effects and need not be independently modeled. For example, there are several policies that lower interest costs for utilities (government-subsidized loans, guaranteed loans, and tax-exempt bond financing). Thus one scenario simulation run with a reduced interest rate represents the effect of all policies that would reduce a utility's borrowing cost.

Selection of policy options was based on:

A review of the literature on the institutional, financial, and regulatory issues concerning electric-utility involvement in solar- and wind-power generating facilities

Interviews (telephone and in person) with utility executives, utility trade associations, and government officials involved in utility wind-energy and related programs [see listing of organizations contacted in the course of this study (table 1–1]

Internal discussions among Arthur D. Little, Inc. staff with experience in utility work

A desire to demonstrate the effectiveness of the methodology

The literature review and interview program provided the background for:

An understanding of the financial criteria used by utilities in making investment decisions

Policy options that, from the perspective of the government and utilities, are feasible, realistic, potentially effective, and are in need of further analysis

1. Higher Return on Equity.

Direct Increase in Allowable Rate of Return on Equity. It can be argued that if a utility's rate of return is fixed, there is no financial incentive to take the risk of investing in wind turbines—any savings from reduced fuel costs would simply be passed through to ratepayers. Thus, a state regulatory commission could grant a utility a higher rate of return on wind turbine investment, as is currently allowed, for example, in California (see chapter 4).

Other changes in existing policies would also have the effect of increasing a utility's return on equity. For example, under PURPA, a utility cannot own more that 50 percent of a small power-production facility and still be a qualified facility. If this provision were eliminated, a utility could own a wind farm and not be subject to the state rate-making process, thereby enabling it to earn a higher rate of return on its wind investment.

To show the effects of any policy allowing a higher return on equity, scenario 3 uses a 16-percent target return on equity for the wind investment only. The remainder of the utility investment remains subject to an 11-percent target rate of return.

Removal of Regulatory Lag. As discussed in chapter 4, regulatory lag during an inflationary period has contributed to the poor financial environment for utilities and has made investment in any new generation equipment more difficult. As a result, the actual rate of return has often been below the target allowed by state regulatory agencies. In order to test the effects of removing regulatory lag, we defined a base case with no regulatory lag (scenario 11) and then added the wind project (scenario 12). To examine the effect of an optimistic case, we doubled the fuel savings resulting from the class 6 simulations and called it the "supercase" (scenario 13).

2. Improved Cash Flow. Cash-flow problems arising from the availability and cost of capital may be a constraint on the ability of a utility to invest in wind turbines even though the investment can be justified by the fuel savings. The following policies were analyzed for their effects on cash flow under the assumption of a constant return on equity.

Investment Tax Credit. Cash flow can be improved by granting a greater investment-tax credit than the 10 percent currently allowed. For example, some of the energy-tax credits that are available to nonutility companies (see chapter 4) under the Crude Oil Windfall Profits Tax Act could be extended to utilities. This would improve cash flow by lowering the first cost of the equipment, thereby reducing borrowing requirements, and, if normalization accounting is allowed, the tax savings could be kept out of the rate base in the year of purchase, thus providing a source of cash in subsequent years. Scenario 4 assumes an additional 15-percent investment tax credit. (It is noted that all the financial simulation herein is based on normalization accounting.)

Accelerated Depreciation. With greater accelerated tax depreciation for wind investment than is allowed under existing law, the utility would write the investment off more rapidly, thus reducing its tax burden and providing a source of funds for future investment. If normalizing accounting is used, the tax savings from accelerated depreciation would not flow through to ratepayers but instead would be deferred. The deferred tax reserve would provide the utility with a source of cash in subsequent years, thereby reducing

Financial and Policy Analysis of Scenarios 131

borrowing requirements. To analyze the effects of accelerated depreciation, we have defined scenario 5, which assumes a five-year tax life, and scenario 6, which assumes a five-year tax and book life.

Construction Work in Progress (CWIP) in the Rate Base. Allowing CWIP in the rate base (see chapter 4 for background discussion) improves a utility's cash flow by reducing the amount of money it has to borrow during the construction period. It allows for rates during the construction period that generate revenues that are sufficient to cover the cost of funds (debt and equity) used to construct a new facility. To analyze CWIP, we define a no-wind base case, assuming rate-base regulation (scenario 14). We then add the wind investment (scenario 15) and finally put CWIP into the rate base (scenario 16).

3. Government Subsidies. Government subsidies and variations thereof are the most frequently discussed incentives in the policy literature on renewable-energy resources. The effect of subsidies is to reduce the overall cost of the investment while also improving the utility's cash flow.

Direct Grants. Direct government grants would lower the cost of wind equipment for the utility and would improve cash flow by reducing the amount of capital that a utility would have to raise. Scenario 7 assumes a 25-percent government subsidy for the purchase of wind turbines.

Loan Guarantees. Government guarantees for utility loans for purchase of wind turbines would improve a utility's access to capital and would result in lower interest cost. This policy would be embodied in scenario 9, which assumes a 7-percent interest rate rather than the 9-percent rate assumed for other scenarios.

Subsidized Loans. Government-subsidized loans to a utility for investment in wind turbines would also lower a utility's interest cost, which is also reflected in scenario 9. A joint-ownership arrangement between a publicly owned and investor-owned utility that can utilize tax-exempt-bond financing is also represented by this scenario.

Reduced Other Taxes. Another form of government subsidy would be to reduce the other (that is, property, and state income taxes.) Scenario 8 assumes a 50-percent reduction in the utility's "other taxes liability."

4. Institutional/Regulatory Changes.

Government Ownership of Wind Turbines (Scenario 10). One way to encourage utility use of wind power is to have the government own the wind turbines and either sell the power or lease the facilities to the utility. The government would incur the capital costs and the utility would not have to incur additional financing burdens. Because of the government's lower borrowing cost and its not having to pay taxes, the cost to the utility would be lower than with the utility investment.

Fuel-Adjustment Clause Modification (Scenarios 17 and 18). The question is whether the fuel-adjustment clause is a disincentive for investment in renewable-energy resources. As discussed in chapter 4, some have asserted that if a utility is allowed to pass its incremental fuel costs through to consumers, it has little incentive to invest in facilities that would save fuel. Thus we construct a base case with no fuel-adjustment clause (scenario 17) and then add the wind investment (scenario 18).

Results of the Financial Model

Criteria

In this section we examine the specific effects of each scenario on the utility's financial condition. Although this generates extensive data on balance sheet, income statement, and cash-flow statements for the utility, we focus on the following indicators of the utility's financial condition:

New common stock

Earnings per share

The ratio of earnings to bond interest (that is, the bond-coverage ratio)

AFDC as a percent of total earnings

Customer rate

We consider these to be the key indicators affecting utility decisions regarding new investment. (See chapter 4 for a discussion of utility-investment criteria.) Return on equity is also shown for each scenario (tables 6–1 and 6–2). However, since we are regulating on the basis of equity with a target return of 11 percent, the variations in return on equity under each scenario should be considered more a result of the model rather than the actual policy effects. For example, in some scenarios the rate of return exceeds the target of 11 percent. Under customary regulatory practice, this higher rate of return would be considered by a regulatory commission in setting the rates for subsequent years so that on the average, the target rate would not be exceeded.

Table 6–1
Comparisons of Measures of Utility Financial Condition with and without Wind Investment

New common stock (10^6^$)

	1985–1989	1990–1994	1995–1999
No wind	0	80	238
Class 4 wind	23	103	211
Class 6 wind	22	103	210
Wind Supercase	23	94	207

Earnings per share ($)

	1985–1989	1990–1994	1995–1999
No wind	5.15	6.20	6.72
Class 4 wind	4.86	6.21	7.05
Class 6 wind	4.88	6.23	7.07
Wind Supercase	4.94	6.28	7.11

Bond-coverage ratio

	1985–1989	1990–1994	1995–1999
No wind	2.49	2.29	2.30
Class 4 wind	2.37	2.35	2.34
Class 6 wind	2.37	2.35	2.34
Wind Supercase	2.39	2.36	2.34

AFDC as a percent of earnings

	1985–1989	1990–1994	1995–1999
No wind	37.8	43.4	45.0
Class 4 wind	40.6	39.4	41.3
Class 6 wind	40.5	39.2	41.4
Wind Supercase	40.2	39.1	41.6

Return on equity (percent)

	1985–1989	1990–1994	1995–1999
No wind	10.7	11.0	10.9
Class 4 wind	10.3	11.4	11.4
Class 6 wind	10.3	11.4	11.4
Wind Supercase	10.4	11.4	11.4

Customer rate (¢/kwh)

	1985–1990	1994–1994	1995–1999
No wind	8.22	10.87	15.05
Class 4 wind	8.21	11.07	15.13
Class 6 wind	8.18	10.98	15.02
Wind Supercase	8.08	10.61	14.61

Table 6-2
Utility Financial Measures for Policy Scenarios

Scenario	New Common Stock ($10⁶)	Earnings per Share ($)	Bond Coverage (Percent)	AFDC as a Percent of Total Earnings	Return on Equity (Percent)	Customer Rate (¢/kwh)
Average for 1985–1989						
(1) No wind	0	5.15	2.49	37.78	10.74	8.22
(2) Class 4 wind	23	4.86	2.37	40.58	10.31	8.21
(3) 16% return on equity	22	4.98	2.40	39.86	10.54	8.24
(4) 25% investment-tax credit	22	4.88	2.37	40.45	10.35	8.21
(5) 5-year tax life	31	4.95	2.31	39.63	10.51	8.14
(6) 5-year tax and book life	18	4.80	2.31	41.18	10.18	8.22
(7) 25% direct subsidy	0	4.98	2.36	40.33	10.42	8.18
(8) Reduced other taxes	22	4.89	2.38	40.33	10.38	8.19
(9) 7% WECS bond	23	4.90	2.41	40.31	10.38	8.20
(10) Government ownership	0	5.18	2.50	37.5	10.8	8.15
(11) No wind, zero lag	0	5.30	2.56	38.75	10.77	8.25
(12) Class 4, zero lag	0	5.26	2.41	40.46	10.70	8.27
(13) Supercase, zero lag	0	5.26	2.41	40.45	10.70	8.13
(14) No wind, rate-base regulation	0	5.10	2.39	41.63	10.64	8.21
(15) Class 4 wind, Rate-Base regulation	24	4.97	2.41	39.96	10.53	8.24
(16) Class 4 wind, CWIP in rate base	18	5.14	2.46	36.52	10.80	8.32
(17) No wind, no fuel-adjustment clause	0	4.88	2.42	39.06	10.25	8.15
(18) Class 4 wind, no fuel-adjustment clause	27	4.68	2.32	41.50	10.0	8.15
Average for 1990–1994						
(1) No wind	80	6.20	2.29	43.43	11.0	10.87
(2) Class 4 wind	103	6.21	2.35	39.4	11.4	11.07
(3) 16% return on equity	91	6.82	2.47	36.2	12.2	11.23
(4) 25% investment-tax credit	100	6.31	2.37	38.9	11.7	11.05
(5) 5-year tax life	105	5.92	2.17	40.1	10.9	10.85
(6) 5-year tax and book life	79	5.99	2.31	41.6	11.0	11.18
(7) 25% direct subsidy	100	6.31	2.37	38.9	11.7	11.05

Financial and Policy Analysis of Scenarios 135

(8) Reduced other taxes	102	6.24	2.35	39.1	11.4	11.01
(9) 7% WECS bond	91	6.33	2.41	35.9	11.5	11.06
(10) Government ownership	76	6.24	2.30	43.4	11.0	10.59
(11) No wind, zero lag	17	6.57	2.33	42.2	11.0	10.90
(12) Class 4 wind, zero lag	57	6.63	2.33	39.4	11.3	11.07
(13) Supercase, zero lag	51	6.64	2.34	39.4	11.3	10.61
(14) No wind, rate base-regulation	81	6.25	2.21	46.7	11.1	010.9
(15) Class 4 wind, rate-base regulation	90	6.60	2.44	36.9	11.9	11.18
(16) Class 4 wind, CWIP in rate base	85	6.77	2.47	35.2	12.1	11.21
(17) No wind, no fuel-adjustment clause	110	4.77	2.00	54.7	8.9	10.57
(18) Class 4 wind, no fuel-adjustment clause	12.2	5.00	2.11	46.5	9.6	10.75

Average for 1995–1999

(1) No wind	238	6.72	2.30	45.0	10.9	15.05
(2) Class 4 wind	211	7.04	2.34	41.3	11.4	15.13
(3) 16% return on equity	202	7.71	2.41	39.4	11.9	15.25
(4) 25% investment-tax credit	188	7.52	2.40	39.8	12.1	15.13
(5) 5-year tax life	218	6.60	2.35	42.8	10.9	15.19
(6) 5-year tax and book life	213	6.98	2.40	43.4	11.3	15.11
(7) 25% direct subsidy	218	6.99	2.33	42.3	11.3	15.01
(8) Reduced other taxes	211	7.07	2.34	41.4	11.4	15.06
(9) 7% WECS bond	45	7.84	2.57	23.3	12.1	15.24
(10) Government ownership	239	6.72	2.30	45.4	10.9	14.65
(11) No wind, zero lag	201	7.71	2.33	45.2	11.1	15.10
(12) Class 4 wind, zero lag	174	7.92	2.34	42.3	11.3	15.13
(13) Supercase, zero lag	170	7.95	2.33	42.6	11.3	14.61
(14) No wind, rate-base regulation	243	6.56	2.18	49.8	10.7	15.11
(15) Class 4 wind, rate-base regulation	202	7.73	2.43	39.0	12.1	15.28
(16) Class 4 wind, CWIP in rate base	203	7.82	2.42	39.1	12.1	15.25
(17) No wind, no fuel-adjustment clause	271	4.62	2.07	55.2	9.1	14.66
(18) Class 4 wind, no fuel-adjustment clause	244	5.26	2.12	49.5	9.7	14.8

The earnings per share is a good measure of the benefit to stockholders of an investment or policy change. Other things being equal, a higher earnings per share will result in higher prices for utility stock. The amount of new common stock the utility must issue is directly related to the earnings per share. Because our synthetic utility has stock that is currently selling below book value, any additional common-stock issue will tend to dilute the earnings resulting in lower earnings per share.

Customer rate shows the combined effects of fuel savings and the revenue needed to meet the additional financing requirements. Within the adversary context of regulatory proceedings, it is usually difficult to justify a policy or investment that results in higher customer rates, even though this is sometimes a desirable action, particularly in the short term.

For simplicity of presentation, we display the scenario results in terms of average performance within each of three five-year periods:

1985–1989

1991–1994

1995–1999

Most of the investment in WECS units occurs during the first period. By grouping the data into five-year periods, we "smooth out" some of the year-to-year variations that inevitably occur when a computer model is used. This smoothing out does reflect what one might expect in actual practice.

The Impact of Wind Investment with No Policy Changes

Table 6–1 shows the results using the criteria just discussed. During the first five years, the utility's financial condition worsens as a result of the WECS investment. Additional common stock must be raised to finance the investment. With the utility's stock selling below book value, the resulting dilution of earnings causes a decline in earnings per share of 29¢ for the class 4 case, 27¢ for the class 6 case, and 23¢ for the supercase. In each wind case, the bond-coverage ratio declines because of the additional interest cost incurred. The AFDC percentage increases because most of the WECS facilities are built and AFDC expenses incurred during this period. The customer rate declines marginally in the class 4 and class 6 case but significantly (14¢) for the supercase because of higher fuel savings. In general, the results for the first five years show that the utility is adversely affected by WECS investment because it is required to obtain additional financing in an environment that is not conducive to new investment. This results in further erosion of the utility's initial financial position.

During the second five-year period, the utility's financial position is somewhat improved. The utility must still issue additional stock ($23 million for the class 4 and class 6 cases and $14 million for the supercase). However, since earnings are higher because of the addition of the wind investment to the rate base, the effect of issuing additional common stock is offset so that the earnings per share remain approximately constant for the class 4 case and increase slightly for the class 6 case and for the supercase. Since few wind turbines are added during this period, little additional financing is required. The bond-coverage ratio increases in each case. The AFDC percentage declines because earnings are higher reflecting an expanded rate base that included the wind turbines, and more of the earnings are from completed plants. While the utility's financial position for the wind cases improves, lower customer rates result only for the supercase. For the two other wind cases, the customer rates are higher because of the higher revenue requirements for an expanded rate base and inadequate offset from fuel savings.

During the last five years, the utility's financial position is improved markedly, primarily because of lower financing requirements. By this period, the rate base is higher due to the wind investment made in prior years. The larger rate base results in higher depreciation charges that provide a source of funds to finance the utility's plant expansion. Consequently, less common stock is issued than for the no-wind case. For each wind case, the earnings per share are significantly higher because of the lower number of shares of new common stock issued and higher earnings from an expanded rate base. The lower financing requirement results from the cumulative depreciation of the earlier investments (recall that these investments increased financing requirements in the years they were made). The bond-coverage ratio increases because the funds generated from additional depreciation result in less borrowing to finance the utility's expansion. The AFDC percentage declines because of higher earnings from an expanded rate base that includes the wind turbines.

The customer rate increases slightly during this last period for the class 4 case, declines slightly for the class 6 case, and declines significantly for the supercase. The increase for class 4 again reflects higher revenue requirements for an expanded rate base, which are not totally offset by reduced fuel costs.

In general, the additional financing costs and the need to provide a fixed rate of return on a higher rate base resulted in a little benefit to the customer from WECS investment, except in the optimistic supercase.

Only the supercase produces fuel savings that are large enough to ultimately reduce customer rates without a significant impairment in the utility's financial condition. But even this case results in increases in new financing requirements during the first five years—a significant obstacle to new investment given the utility's adverse financial environment. The analysis for the supercase also illustrates a basic problem of rate regulation—that is, that savings in operating costs will be passed through to customers in the form of lower rates rather than to benefit the utility directly for the investment.

Project Decision Factors Using Financial Results

Recalling chapter 5, engineering-economic analysis for the assumed conditions concluded:

> A class 4 wind project may be "go" for 100 MW but not likely for anything larger, given current economic and policy assumptions.
>
> A class 6 wind project would be "go" at the full 1000 MW simulated.
>
> A "supercase" in which the savings of the class 6 case are doubled would be a strong "go" at 1000 MW.

Reference to the financial results in table 6–1, in the light of the preceding discussion, suggests that none of the foregoing cases is likely to elicit a "go" decision. Clearly, in today's utility-investment and regulatory environment, it cannot be assumed that an economic wind-energy project will be viewed as a favorable investment.

In running scenarios of policy change to this environment through the financial model, we selected class 4 as the baseline for comparison of results. This is because 3.6 percent (16,000 Km^2) of the land area in the Northeast is class 4 or better, while only 0.11 percent (460 Km^2) is class 6 or better. If wind-energy development is to make a significant impact on this region, it was felt that class 4 sites will have to prove viable for investment. Thus our objective in considering policy changes is to improve both the financial and economic results.

The class 6 land is mainly in the northern sector of the region, with 360 Km^2 in Maine, and 62 Km^2 in New Hampshire and Vermont. The southern portion of the region has few or no class 6 sites, except in the offshore environment. Since our wind data comes from Boone, North Carolina, there is a bias in the simulation toward the southern inland portion of the region, which has a generous endowment of class 4 sites (see table 5–1). It would, of course, be interesting to rerun the models and generate additional data in the form of table 6–2, using class 5 and class 6 sites as the baselines. This proved to be beyond the scope of the study.

The Effects of Policy Changes

Reference is made to table 6–2 in which the financial-simulation results for policy variations are presented for each of five key financial indicators, averaged over each of three five-year periods, spanning 1985–1999. Our interpretation of these results is summarized below.

Higher Return on Equity (Scenario 3). A 16-percent return on equity for the wind investment results in a small improvement. New common stock issued is slightly lower, especially in the second five-year period when it is $12 million lower. This is because some of the additional earnings are retained by the utility for future expansion. Resulting earnings per share are 12¢ higher during the first period, 61¢ during the second period, and 66¢ for the third period. These higher earnings result in a lower AFDC as a percentage of earnings, and a higher bond-coverage ratio. The customer rates are increased for each of the three periods because higher revenue requirements are necessary to meet the 16-percent return on the WECS investment. Thus, ratepayers are, in effect, subsidizing the utility's investment in WECS.

Although a higher return on equity for the wind investment improves the utility's performance in the first five years, it does not significantly alter the fundamental attractiveness of investment in WECS when compared to the no-wind base (scenario 1). In general, the effects of the policy change at 16 percent are too small to overcome the overall financial effects of the WECS investment in the first five years but are effective thereafter. Perhaps a higher return, say 20–24 percent can be considered for the first five years and then reduced later.

Investment-Tax Credit (Scenario 4). A 25-percent investment-tax credit (compared to 15 percent in all other scenarios) has little discernible effect on the utility during the first five years. Assuming the use of normalization accounting, all that changes is the amount of tax the utility would have to pay. Customer rates, new financing requirements, and the other financial indicators, would change very little.

Because of the assumption of normalization accounting, the beneficial effects of a tax credit are seen in the two periods between 1990 and 1999. Less common stock is issued because the revenue from the deferred tax reserve becomes a source of financing. Less common stock results in a significant increase in earnings per share. There is little change in the customer rate during these two periods. There is a small increase in the bond-coverage ratio and a small decline in AFDC as a percentage of total earnings.

Five-Year Tax Life (Scenario 5). A five-year tax life appears to have a negative effect on the utility. The new common stock required is higher for each five-year period because of deferral of investment-tax credits. The tax laws limit the investment-tax credit to 90 percent of the income-tax liability. Since the utility was at the 90-percent level before the tax life was reduced, the investment-tax credit was also reduced. The result was that the investment-tax credit, which would normally be used to offset new financing by the utility, is instead carried forward until it can be used.

Five-Year Tax and Book Life (Scenario 6). A policy that includes reduced tax and book life for WECS units significantly improves the cash flow of the utility during the first two five-year periods. Less common stock is issued because the higher depreciation expense from the WECS unit results in more cash being available for the utility's capacity expansion. New common-stock requirements are slightly higher in the last period because most of the depreciation will have already been taken. Despite the lower common-stock requirement, earnings per share decrease for each period because the shortened depreciation period prevents the utility from taking the full investment tax credit (see explanation under Five-Year Tax Life Case), resulting in slightly higher taxes and, consequently, lower earnings. There is little change in the customer rate in either period.

Direct Federal Subsidy (Scenario 7). A 25-percent direct federal subsidy improves the ability of the utility to invest in WECS by reducing the new financing that is required. The effect of the subsidy in the first five years is to reduce significantly the new common stock that must be issued by $23 million. Lower common-stock requirements for the investment results in less dilution of earnings, yielding a 12¢ increase per share. The effects on the AFDC percentage and on the interest-coverage ratio are marginally beneficial. The lower financing requirements also result in some small decline in the customer rate.

The beneficial effects of a direct subsidy are evident during the first five-year period when most of the subsidy is received. Thus, for the last two five-year periods, the effects of the subsidy policy are small. The new common-stock requirement is somewhat higher because the lower cost for WECS units results in less depreciation as a source of funds to finance the utility's capital expansion. Higher common-stock requirements result in significantly lower earnings per share during the final period. The effects on the other indicators are minor.

Reduced Other Taxes (Scenario 8). Reducing the nonfederal taxes paid by the utility has little effect. The customer rate declines slightly, as the tax savings are passed through to ratepayers. There is little effect on common-stock financing or on earnings per share. In general, the relatively small size of the wind investment compared to the total system and the insignificance of nonfederal taxes compared to other financial items, results in minor financial improvement. [Note, however, that in the project analysis, the effect is to decrease FCR by 0.25 (see figure 6–1) resulting in a significant increase in breakeven cost. This underscores the need to study the impact of an incentive on the utility as a whole since the internal project analysis can overstate the effects of an incentive.]

Seven-Percent WECS Bond (Scenario 9). A lower interest rate for WECS units has little change on the utility during the first five years. The customer rate, new common-stock requirement, and AFDC percentage change very little. The interest-coverage ratio improves somewhat because the lower interest rate reduces interest charges. Earning per share appear to improve slightly because of lower interest cost.

During the last two five-year periods, the effects of a lower interest rate are dramatic. New common stock issued declines from $103 to $91 million in the second period and drops to $45 million in the final period. The lower common-stock requirement results in a 12¢ earnings per share increase in the second period, and a 79¢ earnings per share increase in the final period. The lower interest rate increases the debt capacity of the utility, thereby reducing the common stock that must be issued.

The lower interest cost also results in a higher bond-coverage ratio in the last two five-year periods. The AFDC percentage declines dramatically in the final period, reflecting an assumption in the model that the AFDC debt component rate equals one-half of the utility's borrowing rate.

Government Ownership (Scenario 10) and Third Parties. In this case the government produces electricity without paying taxes and at a minimal financial cost of capital. This allows the utility to purchase wind-generated electricity at a price considerably less than the avoided cost, without making any investment. For the first two periods, scenario 10 is a clear improvement over scenario 2 in which the utility invests. By the third period, however, earnings per share would have increased in scenario 2 but not in scenario 10. Common-stock requirements are higher in the third period, and earnings per share are lower because less depreciation is available to finance the utility's expansion.

There are government entities, like the Western Area Power Authority, which can develop wind power for sale to utilities under conditions analogous to scenario 10. We note that virtually all operating cost savings to the utility are passed onto the customer, and there is virtually no benefit (and no risk) to the investor. While this is a no-risk scenario for the utility, it does not serve to motivate the utility itself to invest in wind power. To the contrary, it constitutes a disincentive.

It is also noteworthy that using class 4 sites, a government entity can produce power at less than the utility's avoided cost. This means that a nongovernment third party, enjoying the same tax exemptions and interest rates, could charge at the utility's avoided cost and make a profit.

We have not studied the implications here for third parties operating under current or proposed financial and tax incentives. Any private third-party entity is assumed to charge the avoided cost no matter what its profit so that there is no financial effect on the utility.

Removal of Regulatory Lag (Scenarios 11–13). The removal of regulatory lag significantly improves the financial condition of the utility. Comparison of scenario 2 and scenario 12 shows marked improvement in the utility's earnings per share. (Note that the model does not provide strict comparability because the base cases to which each class 4 case is referred are different. With regulatory lag, the ROE is an outcome of the model. Without regulatory lag, the ROE is fixed by the input.) With removal of the lag, the WECS investment enters the rate base a year earlier, resulting in higher earnings. Additional funds generated from depreciation of the larger rate base lowers the amount of common stock that must be issued to finance the utility's planned capacity expansion. The combination of a lower common-stock requirement, earlier recovery of inflation increases, and higher earnings results in a significant increase in earnings per share—that is, 40¢ for the first period, 42¢ for the second period, and 87¢ for the third period. The effects on the bond-coverage ratio and on the AFDC percentage of total earnings are small. The customer rate is only affected by the removal of regulatory lag in the first five years (0.06¢ increase) during which time the WECS units enter the rate base. It appears that in later years, the higher depreciation charges and other adverse effects on the customer rate of removing regulatory lag were compensated by earlier fuel savings that accrue a year earlier.

It is in the supercase where the effects of removing the regulatory lag are most noticeable with regard to customer rates. Comparison of scenario 11 and scenario 13 shows a small effect on earnings per share, but the customer rate declines significantly because the fuel savings are passed through to the customer one year earlier.

The pure effects of removal of regulatory lag can also be seen by comparing the no-wind base cases of scenario 1 and scenario 11. Clearly, removal of regulatory lag alone benefits the utility. Since the utility's investment in capacity expansion enters the rate base one year earlier, the additional depreciation generated each year becomes a source of funds for additional investment, thereby reducing the amount of common stock that must be issued.

Less common stock and higher earnings from an expanded rate base result in a higher earnings per share. The effects on the bond-coverage ratio and on the AFDC percentage are minor. Customer rates go up slightly, as expected: (1) to allow the utility to recover its costs one year earlier, and (2) because revenue requirements are higher because of the expanded rate base.

It should be noted that one reason that removal of regulatory lag appears to be one of the more effective measures for improving the utility's fiscal position is that its effects are not limited to the wind investment. Most of the other incentives considered here (for example, CWIP in the rate base) apply only to the wind investment, so incremental effects on the overall system are small. Removal of regulatory lag, however, is applied to the entire utility system and improves the overall investment climate.

CWIP in the Rate Base (Scenarios 14–16). The effects of including CWIP in the rate base for WECS can be seen by comparing scenario 16 and scenario 15. The results are beneficial to the utility—primarily during the first five years, during which most of the wind-turbine construction takes place and when the corresponding CWIP charges are included in the rate. New common stock is lower for the first two five-year periods because the CWIP revenue becomes a source of financing new investment. It increases in the last five-year period because the revenue from CWIP charges, which would have been available if CWIP were not in the rate base, has been utilized earlier. The lower common-stock requirement results in significantly higher earnings per share during the first two five-year periods.

The bond-coverage ratio and the AFDC percentage follows the same pattern—improving in the first two five-year periods because of lower financing requirements, and slightly increasing in the last period because the revenue from CWIP charges has already been received.

As would be expected, including CWIP in the rate base results in significantly higher customer rates during the first five-year period, as current ratepayers pay the CWIP charges. In the second five-year period, the effects are small. For the last five years, the rate declines somewhat because the CWIP charges would have already been included in the rates for the first ten years, thus reducing the amount that has to be recovered from the rates in the last five years.

Elimination of the Fuel-Adjustment Clause (Scenarios 17 and 18). Elimination of the fuel-adjustment clause clearly has negative effects on the utility. For example, in comparing scenario 17 with 1 and scenario 18 with 2, respectively, the customer rate declines during each period as the increasing fuel costs become subject to regulatory lag, rather than being recovered in the year in which they are incurred. The common-stock requirement increases because the lower utility revenues result in less internal funds being available to finance new investment. As a result of higher common-stock requirements and lower earnings, earnings per share drop dramatically. The bond-coverage ratio also shows major negative effects because of lower revenues and higher interest costs.

The reasons for these adverse effects are intuitively obvious. By not allowing timely recovery of higher fuel costs, the utility is forced to recover them through the rate-making process, thereby accentuating the adverse financial effects of regulatory lag.

Granting a regulatory climate in which there is no fuel-adjustment clause, we can compare scenario 18 with 17. In the first five years the effect of the wind project is to increase common-stock issues, reduce earnings per share by 20¢ (compared with 29¢ in scenarios 1-2), reduce bond coverage and increase percentage AFDC, while customer rate is unchanged. By the second five-year period, there is a 23¢ increase in earnings per share (compared with only 1¢ from scenarios 1 to 2), but customer rate has increased by 0.18¢/kWh (compared with 0.20¢/kWh from scenarios 1-2). By the third period, earnings improved by 60¢ (compared with 33¢ in scenarios 1-2), but customer rates have increased by 0.14¢/kWh (compared with 0.08¢/kWh in scenarios 1-2).

While the introduction of the wind project in the absence of the fuel-adjustment clause does produce a measureable improvement in financial impact, it is not sufficient to compensate for the basic adverse effects imposed by loss of the fuel-adjustment clause.

Adverse effects are, of course, expected when the fuel-adjustment clause is removed. A significant policy question would be whether use of the fuel-adjustment clause should be limited to companies that demonstrate an investment into fuel conservation. This is a complex issue beyond the scope of this study.

Identification of Effective Policies

The most effective policies are considered to be those that reduce the need to issue new stock in order to finance the wind investment and at the same time minimize the cost to the ratepayer. The model results show the following to meet these criteria:

Direct subsidy (scenario 7)

Increased investment-tax credit (scenario 4)

Reduced bond interest (scenario 9)

The direct subsidy is most effective during the first five years. The investment-tax credit is felt in the second and third five-year periods. The reduced-loan rate is effective in all periods. These three scenarios all require the introduction of federal-incentive programs.

Financial and Policy Analysis of Scenarios

In scenario 2, most of the adverse effects of the wind investment on the utility occur in the first five years. Thus incentives that provide financial relief during this period are considered to be particularly effective. These include:

Zero regulatory lag (scenario 12)

CWIP in rate base (scenario 16)

Increased return equity (scenario 3)

Zero regulatory lag is the most effective, as it represents a major structural change to the regulatory environment. These policies are effective in all three periods. None of them requires federal action, and can be individually applied at the state level by legislatures for implementation by the regulatory commissions. However, they all result in varying increases to customer rate. In all cases, improvement of the utility's financial posture is at the expense of the customer.

It is noted that none of these policies fully restores the utility to its base-case financial condition, they merely ameliorate the adverse effects to different degrees. Some combination of effective policies may be desirable, but it was beyond our scope to develop this.

Policies considered not to be effective, based on our analysis are:

Reduced tax life (Scenario 5)

Reduced tax and book life (Scenario 6)

Reduced other taxes (Scenario 8)

Removal of fuel-adjustment clause (Scenario 18)

To some extent, the ineffectiveness of reduced-tax and book-life scenarios was a circumstance of the utility's tax situation, but this is considered to be a common or typical circumstance for electric utilities today. Reduction of the "other" nonfederal taxes represents a considerable amount of revenue, but most of the benefit is shifted to the customer as reduced rate.

Third-party arrangements are not an incentive for the utility to invest. The impact of any third-party arrangement whereby the utility pays the avoided cost by definition is completely neutral. The impact of government as the third party (scenario 10) is positive because the customers enjoy a lower rate, and the utility makes no new investment. It may be useful to consider municipal-utility partnerships to take advantage of a tax-free, reduced-interest opportunity while asserting a utility stake in the outcome.

Implementation of any of the policies identified as effective would need to be supported by considerations beyond the scope of this report. Federal subsidies must be shown to yield a national benefit. Schemes that increase cost to the customer need to be rationalized on equity grounds and successfully defended in an adversary and highly politicized environment. These and related issues are discussed in the following section.

Overriding Public-Policy Considerations

Most of the incentive measures studied here involve many regulatory-policy concerns that currently confront utilities. While the incentives have been described within the context of these issues, there are overriding public-policy objectives and political considerations that transcend the question of utility incentives for wind, as discussed in the following paragraphs.

Elimination of Regulatory Lag. Elimination of regulatory lag was shown to have very salutary effects on the utility's financial condition; however, regulatory lag is an issue that has confronted the utilities for some time and is, in part, responsible for their current poor fiscal condition. State regulatory agencies have the authority to mitigate the adverse effects of regulatory lag, but these agencies have varied in the extent to which they have done so. Thus, while our analysis has cast the regulatory-lag issue in terms of an incentive for wind-turbine investment, it must be recognized that the issue has a broader context within which it will be considered in the policy-making process. Nevertheless, resolution of these issues will have a direct bearing on the financial condition of utilities and consequently their fiscal capacity to invest in fuel-savings technologies.

CWIP in Rate Base. While it improves the utility's fiscal condition by providing a source of funds for new investment, issues of an economic, political, and philosophic nature surround the question of whether CWIP should be included in the rate base. We have analyzed a policy whereby only the wind-turbine CWIP would be included in the rate base. The assumption in doing so is that this differential treatment is an incentive for the utility to invest in wind turbines. It can also be shown, however, that allowing CWIP in the rate base for *all* the utility's investments would improve the utility's cash flow to the extent that it would improve its fiscal capacity to invest in wind turbines. Thus it may be that policies that improve the utility's financial condition overall, such as removal of regulatory lag and including CWIP in the rate base, may be more effective than more specific policies targeted toward wind turbines. Since these policies result in increased customer rates, the issue will remain controversial.

Fuel-Adjustment Clause. The fuel-adjustment clause has become increasingly controversial as it is a clearly identifiable item contributing to the increase in consumer bills. Our analysis shows that elimination of the fuel-adjustment clause is very damaging to the utility. Nevertheless, it will continue to be argued by some (see chapter 4) that increased fuel cost pass-through denies the utility an incentive to utilize fuel more efficiently or to invest in capital facilities to reduce fuel use. Thus it is argued that the fuel-adjustment clause should be eliminated, or at least modified, to make it less automatic. This issue arises within the context of the responsibility of state regulatory agencies to evaluate whether utilities are using fuel efficiently and are attempting to realize the renewable resource and conservation investment goals that regulatory agencies have set for them. The extent to which achievement of these goals may be related to regulatory leverage in manipulating a fuel-adjustment clause will also remain controversial.

Increased Return on Equity. If the riskiness of wind turbines is an obstacle to investment, an offer of a higher rate of return may be sufficient incentive to overcome that risk. This opens the question of equity—who should pay for the risk, who should benefit, and who makes the decision about the risk?

In situations where lower customer rates are not achieved in the long term, the equity of a higher-rate-of-return policy may be questionable. The result is that ratepayers simply subsidize a higher rate of return for stockholders. The nation as a whole may benefit from less use of imported fuel oil and the stockholders benefit from a higher return—but the cost is incurred by the ratepayer. Whether ratepayers should bear the cost of a policy that benefits others is an equity issue that will continue to be controversial.

Federal Incentives. The federal-incentive policies raise less complex policy issues. They involve costs to the federal treasury that are readily ascertainable, including:

The direct cost of a subsidy from the treasury

Tax dollars lost as a result of an investment-tax credit

The cost of the treasury of subsidizing a lower loan rate for the purchase of wind turbines

The time-value-of-money costs from accelerated depreciation

Administrative costs

In evaluating their desirability, the cost of alternative incentive measures can be weighed against the benefits derived from reduced use of fuel oil, and the extent to which the incentives are effective in influencing utility behavior.

One question our analysis implicitly raises is: Why not give utilities the same federal incentives as nonutilities for investment in wind turbines? Our analysis shows the potential effectiveness of tax credits and higher rates of return in improving the prospects for wind-equipment investment. These, however, are precisely the incentives that are available to non-utility third parties under PURPA and the Crude Oil Windfall Profits Tax Act. Clearly, the present policy environment is more conducive to third party than to utility investment in wind. If the public-policy objective is to decrease our reliance on imported oil by encourageing use of renewable-energy resources, then in the absence of some overriding policy objective to the contrary, some means of granting utilities incentives similar to those available to third parties appears to make sense. In fact, it can be argued that utility customers would be better off with utility rather than third-party investment in wind turbines because of their experience in generating power and their high standard of equipment maintenance.

Under present policy, utilities have little incentive to invest in wind. The riskiness of the investment is not compensated by financial incentives to the utility investor. By motivating only third parties, the present policy may discourage economies of scale in wind farming, and technological learning on the part of utilities may be impeded. There is ample federal as well as industry precedent for direct utility involvement in acceptance and implementation of new generating technology—for example, through demonstration programs and subsidies as well as state regulatory-commission incentives in the form of allowance for research-and-development costs in rates.

Public-Owned Utilities. This study has focused on a typical "synthetic" private-investor-owned utility. However, our analyses illustrates that using standard utility financial analyses, any investment that is capital intensive is more attractive to a publicly owned utility, primarily because of public policy (tax-exempt financing) that gives a lower cost of capital and tax exemptions that reduce operating costs. Thus, if an objective of public policy is to encourage investor-owned utilities to invest in capital-intensive fuel-savings facilities, certain obvious policy choices flow from the comparison. We have considered the possibility of giving private utilities access to the tax-exempt financing available to publicly owned utilities by using a joint-ownership arrangement. It should be noted, however, that if a suitable arrangement could be devised, ratepayers would benefit from the low-cost financing and from the expertise of the private utility in running a large power-generation system. Our analysis suggests that such innovative institutional arrangements could act as an incentive.

Strategic Considerations. The changing economics of wind-turbine investment over time suggest that incentives should be examined in a strategic manner, in relation to the time periods during which they are needed, and the ease in which they can be eliminated when no longer required. Thus, as more wind turbines are produced, the price may decline to levels where, for example, a subsidy may no longer be needed. Eventually the confidence in

Financial and Policy Analysis of Scenarios 149

wind turbines may reach a point where a capacity credit can be accepted by the utility planner. This would dramatically improve the impact of the investment on the utility. In general, the timing of incentives should coincide with the appropriate stage in the commercialization process of the technology so that: (1) The most effective incentives can be utililized at the stage they are needed, and (2) The incentives can be terminated when they are no longer needed.

7

Summary of Major Findings

Utility Investment in Wind-Power Systems

1. Large wind-energy projects in the multi-megawatt range will be economic for many utilities in the 1980s if wind-industry production-cost targets are met. Several American companies are actively involved in commercial development or large wind turbines, and integration with the gird is being demonstrated.

2. Given the generally adverse financial condition of the electric-utility industry today, the climate is poor for investment in general and for projects perceived as "risky" such as wind energy.

3. Utilities are risk averse with respect to renewable-energy investments. They perceive the benefits of a profitable investment as limited because savings in operating costs are passed on to the customer. They perceive the risks as unlimited because the cost of a poor investment must be borne by the investors.

4. Given the intermittent nature of the wind resource and the inherent conservatism of utilities, wind projects must justify themselves on the basis of fuel-savings only. They are generally not accepted by utilities as a substitute for other capacity in utility-expansion plans.

5. Any investment into renewable energy must meet criteria related to both the economic performance of the project and to the impact on the utility's overall financial performance. Economically sound projects will be rejected if they have an adverse financial impact, even in the short run. The sensitivity on this point is heightened in an era when the electric-utility industry is in a general state of financial distress.

6. Government entities that pay no taxes and enjoy minimal financial costs and long write-off periods can market energy from the wind project studied herein at considerably less than the utility's avoided cost. Thus, in those regions where government-power-marketing entities, such as the Western Area Power Authority, invest in wind power, utilities may prefer to purchase the power rather than invest in their own facilities.

7. Current policies that give incentives to third-party power producers to invest in wind, while exempting electrical utilities, constitute a definite disincentive to the electric-utility industry.

Investment-Evaluation Criteria Used by Utilities

1. A utility will invest in new equipment in order to: (1) maintain a target systemwide reliability as the load grows; (2) obtain long-term efficiencies in operations and maintenance. (Investments to replace retired equipment generally accomplish both objectives. Investments whose purpose is to conserve fuel contribute to reduced operating cost.) A utility will also invest in response to regulatory requirements related to environmental protection or restrictions on the use of particular fuels.

2. Utility investments are subject to two categories of evaluation: (1) internal project criteria using the tools of engineering-economics; (2) financial criteria related to the effect of the project on the overall financial performance of the utility.

3. The internal project criteria include: ratio of the present value of future benefits to costs, internal rate of return on the investment, levelized busbar cost of energy produced, number of years to recover the investment, and the capital cost that can be recovered by the future benefits (that is, breakeven cost). The latter two are frequently used to evaluate investments into fuel-conservation technologies such as wind power.

4. The most significant financial performance criteria for evaluating utility investments are:

New common-stock requirements

Bond coverage

Allowance for Funds Used During Construction (AFUDC) as a percent of earnings

Return on equity

Customer rate

5. Common stock issued at a price below its book value to finance major investments dilutes the equity of the utility and risks a further lowering of market/book ratio. This further limits the ability of the utility to raise equity capital as the value of its shares declines.

6. Lowering of bond coverage may result in an adverse financial rating, increasing the cost of bonds, and sometimes jeopardizing the company's ability to raise money in the bond market.

7. The other financial criteria are important but less critical. A utility seeks to minimize the percentage of AFUDC as earnings. It also seeks to minimize the customer rate while maintaining a target return on equity.

8. If public-policy incentives for wind energy are to be effective, they must have a demonstrable positive effect on the financial performance of the utility in the long term and a minimal adverse effect in short term.

Summary of Major Findings

9. Computer tools of analysis are available to perform investment analysis for electric utilities. These include production-cost models, capacity expansion models, and financial models, as well as simulation of the production of intermittent energy such as wind. This report demonstrates that these tools can be effectively deployed in an integrated manner at a reasonable cost to study the effects of government policy changes on the utility's investment in fuel conserving technology such as wind power.

10. We employed a case-study technique using selected computer methods of analysis to measure the "effectiveness" of policies, mainly using the financial criteria. *Effective* in this context means that the adverse financial consequences of an investment in a 1000-MW wind-energy project are significantly ameliorated by the policy. None of the policies tested, of themselves, fully restored the utility studied to its base-case (that is, no wind investment) financial condition in the short run (first five years).

11. Formulation of effective government policies, as well as of wind-energy promotion and marketing policies, requires that the financial conditions of the industry and the impact of the policy on that condition be fully taken into account.

12. Most of the literature reviewed on incentive policies for wind energy concentrates on internal project criteria such as breakeven cost and levelized busbar cost. It does not take into account the effect of proposed incentives on the financial condition of the utility.

The Effects of Public-Policy Incentives

1. Policies shown to constitute the most effective incentives in our case study include:

Removal or reduction of the adverse effects of regulatory lag

Allowance of construction work in progress (CWIP) for the wind project in the rate base

Increasing the allowable rate of return for the wind investment, by at least 4 percentage points

These policies involve increases to the customer rate and need to be justified by arguments related to equity. They can be implemented at the state level.

2. Policies shown to simultaneously benefit the investor and the rate-payer are:

Direct grants or subsidies

Reduced bond interest rate for wind financing

Increased investment-tax credit for wind

These involve costs to the government and need to be justified by arguments related to national benefits.

3. Policies shown to be ineffective in our case study include:

Accelerated depreciation as reduced tax life

Reduced tax and book life

Reduced nonfederal taxes

Removal of fuel-adjustment clause

4. The ineffectiveness of accelerated depreciation results from a limit on the total allowable depreciation and investment-tax credit. (The impact of the Economic Recovery Tax Act of 1981 was not measured in this study.) Simply stated, the utility in our case study does not pay sufficient taxes to benefit from the accelerated-depreciation scheme. This effect may be less pronounced or eliminated in a financially more viable company. In that circumstance, accelerated depreciation may prove to be an effective incentive.

5. Removal of the fuel-adjustment clause makes any capital investment more difficult. Any improvement in the financial performance of the wind project is more than offset by the adverse effects of removal. A significant policy question is whether use of the fuel-adjustment clause should be limited to companies that invest in fuel conservation. This is a complex issue beyond the scope of study.

6. The finding that certain incentive policies are "effective" or "ineffective" in our case study should not be construed as an advocacy position. There are overriding public-policy considerations related to equity and national benefits that are discussed in this report but are not explored in depth. Our major intent has been to use systems analysis to raise the level of understanding that should support any enlightened debate on public policy in the electric-utility area.

7. The policy findings reported are valid in today's electric-utility financial and regulatory environment. Some of these policies may be less effective in a financially healthier industry and, for that matter, less needed.

8. All of the preceding conclusions are presaged on the assumption that wind-energy technology will perform in a satisfactory manner at an appropriate level of cost. The progress in wind-energy technology to date is in no small measure to the credit of federally funded programs of research, development, and demonstration. Actually bringing utility-scale wind energy into the marketplace in a major way will require continuous efforts to produce more reliable and cost-effective systems.

Index

Accelerated depreciation, 41, 73–74, 75, 130–131, 154
Accounting procedures, 74–75
Allen-Warner Valley coal plant, 42
"Allowance for funds used during construction" (AFUDC), 71–72, 152
Alukarleby, Sweden, 22
Aluminum Company of America (ALCOA), 14–16
Averch-Johnson effect, 73
"Avoided costs," 80–81

"Base case," 93, 128
Battelle Wind Atlas, 94
Bechtel International, 18–19
Bendix Corporation, 17
Benson, C.C., 21, 57
Biomass/tidal/wave-energy conversion, 7
Boeing Engineering and Construction Company, 12, 14, 17–18
Bond ratings, 63–64, 115, 141, 144, 152, 153
Bonneville Power Administration (BPA), 12
Boone, N.C., 96–100
Boyd, D.W., et al., 41
"Breakeven cost," 35–36, 118–123, 152
British Aerospace, 24
Bushland, Tex., 16

California, 72, 81, 84–88
California Institute of Technology, 36
Canada, 25
Capacity credit, 38–40, 126
Capital investment, 12, 57, 127; and technological innovation, 58–60
Carter administration, 77
Cash-flow improvements, 130–131
Central Lincoln Public Utility District (Oreg.), 16
Coal power plants, 1

Coal prices, 106, 107, 108
Cogeneration, 7, 78, 85
Colorado River Storage Project, 17
Computer input, 114–115, 153
Construction work in progress (CWIP), 71–72, 131, 143, 145, 146, 153
Crude Oil Windfall Profits Tax Act (1980), 84, 130, 148
Customer rate, 136, 137, 152
Cuttyhunk Island, Mass., 17

DAF-Indal Corporation (Canada), 25
Danalith Ltd., 22
Darrieus wind turbine, 9, 14
Denmark, 21–22
Distillate-oil price, 106, 107, 108
"Double-dipping" ban, 84

Electric Power Research Institute (EPRI), 43, 100
Electrical Research Association (UK), 24
ELFIN, 42–43
"Emergency rate relief," 115–118
"Energy property," 83
Energy Tax Act (1978), 83–84
Engineering-economic analysis, 93, 112–114, 118–123, 138
Environmental Defense Fund (EDF), 39, 42–43
Environmental-impact statements, 75–76
ERNO Raumfahrtechnik (West Germany), 23
Eugene (Oreg.) Water and Electric Board, 14–16
Expansion plans, 103–104, 127

Fairfield, Calif., 20
Federal Aviation Administration, 75
Federal Communications Commission, 75

Federal Energy Regulatory Commission
  (FERC), 78, 80
Federal Power Act, 20, 80
Felak, R., 42
Financial indicators, 132–136
Financial-simulation model, 42–43, 49–
  51, 93, 112–118, 139–144
Financial statistics, 63–68
Fixed-charge rate (FCR), 35–36, 118–
  120
Flow-through accounting, 74–75
Fuel-adjustment clause, automatic
  (AFA), 58, 69–70, 77, 132, 143–
  144, 145, 146–147, 154
Fuel costs, 1, 57, 105–108, 114, 120–
  121
Fuel displacement, 110–112
Fuel-savings methodology, 32–35, 119–
  120, 151

Garate, J. A., 40
GEC Power Engineering Ltd. (UK), 24
Gedser, Denmark, 22
General Electric Company (GE), 33,
  42, 114; Electric Utility Systems
  Engineering Department
  (GE/EUSED), 47; Financial
  Simulation Program (FSP), 49–51;
  Monthly Production Simulation
  Program (MPS), 47, 109
Generation mix, 103–104
Geothermal energy, 7
Goodnoe Hills, Wash., 12, 18
Gotland, Sweden, 22
Government ownership, 131–132, 141–
  142, 145, 151
Grants, direct, 131, 153
Grosse-Windenergie-Analage
  (Growian) (West Germany), 23

Hamilton Standard Division of United
  Technologies, 17, 18–19, 23
Hansen, D.B., 57
Hawaiian Electric Company (HECO),
  18–19, 81
Heliostats, 8

Horizontal-axis wind turbines
  (HAWTs), 9
Hunt, V.D., 22
Hydro-Quebec, 25
Hydroelectric power, 7

Incremental cost, defined, 80
Inflation, 114
Installed cost, 122–123
Intermittency, 7
Internal project criteria, 152–153
Investment climate, 125–126, 151
Investment criteria, 132–136, 151, 152–
  153
Investment incentives, 4, 56–57, 58–60,
  61–63, 82, 121–122; "effectiveness"
  of, 153–154; federal, 147–148;
  timing of, 148–149
Investment return (ROI), 57, 76–77,
  86–88, 151, 153. *See also* Return
  on equity
Investment-tax credit, 41, 56, 75, 83–
  84, 130, 139, 144, 154

JBF Scientific Corporation, 33, 38
Johanson, E. E., 36, 38
Johanson, E. E., et al., 28
Johanson, E. E., and Goldenblatt, M.,
  38
Justus, C.G., and Hargraves, W.R., 45
Justus, C.G., and Mikhail, A.S., 45

Kalkugnen, Sweden, 22
Karlskronavarvet AB (Sweden), 19, 23
Karlstads Mekaniska Werkstat AB
  (KMW) (Sweden), 23
Klickitat County (Wash.) Public Utility
  District, 12
Korber, F., and Thiele, H.A., 23

Lerner, J.I., 61
"Levelized cost," 35–36, 152
Life-cycle model, 35–38, 118
Lindley, C.A., and Melton, W.C., 33,
  39, 42
Lindquist, O.H., and Malver, F.S., 40
Little Equinox Mountain, Vt., 16

Index

Load characteristics, 103
Loan guarantees, 131
Loan, subsidized, 131, 147
"Loss-of-load-probability" (LOLP), 38
Lotker, M., et al., 41
Lundsager, P., et al., 22

Magdalen Islands, Canada, 25
Maglarp, Sweden, 23
Maine, 138
Malmo, Sweden, 23
Martha's Vineyard, Mass., 16
Medicine Bow, Wyo., 17, 40
Mehrkam/Energy Development Company, 16–17
Metropolitan Edison (Hamburg, Pa.), 16
Minnesota Power and Light, 40
Mississippi, 81
MOD wind turbines, 10–12
Monopoly, 55–57

NASA: Lewis Research Center, 9; Plumbrook facility (Sandusky, Ohio), 10
National Energy Act (1978), 60, 78
National Research Council (NRC) (Canada), 25
NEPA, 76
New England Gas and Electric Association (NEGEA) (Mass.), 38
New England Utilities, 1
New Hampshire, 138
New York Power Pool, 39
Nibe, Denmark, 22
Normalized accounting, 74–75, 130
North of Scotland Hydro Electric Board (UK), 24
Northwest Territories, Canada, 25
Nuclear power plants, 1, 58–59, 105

Oahu, Hawaii, 18, 39, 79
Ocean-thermal-electric conversion (OTEC), 7
Off-shore wind-energy projects, 25–28
Oil and natural gas restrictions, 82–83

Oland, Sweden, 22
Orkney Islands, Scotland, 24

Pacific Gas and Electric (PG&E) Company (Calif.), 12–14, 20, 39, 42, 61, 85–86
Pacific Power and Light (PP&L), 17
Papay, L.T., 62
Pederson, B.M., and Nielsen, P., 22
Pennsylvania, 94
Percival, D., and Harper, J., 33–35
Photovoltaic (PV) electricity generation, 7, 8–9
Policy analysis, 41, 121–122
Policy scenarios, 128–129
Powerplant and Industrial Fuel Use Act (1978), 82–83
Present-value method, 36
"Primary energy source," 79
Prince Edward Island, Canada, 25
Production-cost model, 47–49, 109–112
"Project Aeolus," 25
Public-owned utilities, 148
Public Utilities Fortnightly, 72
Public Utility Holding Company Act (PUHCA), 20, 79–80
Public Utility Regulatory Policies Act of 1978 (PURPA), 20, 21, 56, 69, 77–82, 148

Ranking engine, 7
Rate-making process, 57–58, 115–118
Reagan administration, 77
Regional wind overview, 94–96
Regulation exemptions, 79–80
Regulatory approval requirements, 75–76
Regulatory lag, 76–77, 126–127, 130, 142–143, 145, 146, 153
Research, development, and demonstrations (federal), 9–14, 83, 127
Research Association of the Danish Electricity Supply Undertakings (DEFU), 22
Residual-oil price, 106, 107, 108

Return on equity (ROE), 77, 115–118, 129–130, 139, 145, 147, 152
Revenue-requirements methodology, 41
Risk aversion, 58–59, 60–61, 147
Rocky Flats, Colo., 16

Saab-Scania Company (Sweden), 22
Sandia Laboratories, 9
San Gorgonio Pass, Calif., 17
Scale economies, 1, 56–57
Simulation model, 43–51, 93, 109–114, 153
Site-approval process, 75, 76
Site simulation, 96–100, 127
Skane, Sweden, 22
"Small power producers." *See* Third-party power producers
Solano County, Calif., 20
Solar-electric power generation, 2, 7
Solar-power satellites, 7
"Solar property," 84
Solar-thermal electricity conversion (STEC), 7, 8
Southern California Edison (SCE) Company, 16, 17, 19, 39, 42, 62, 81, 86
Spinning-reserve criterion, 48
State regulatory commissions, 80–81
Stirling engine, 7
Stobaugh, R., and Yergin, D., 8
Stock dividends, 63–66, 118, 141, 152
Stone and Webster Engineering Corporation, 41
Storage, 7, 32, 40–41
Subsidies, federal, 83, 131, 140, 144, 147, 153
Sullivan, W.N., 14, 16
Sweden, 22–23
Swett Ranch, Solano Co., Calif., 20
"Synthetic utility," 31, 43, 93, 100–108

Tax/book life, reduced, 140, 145, 154
Tax Reduction Act (1975), 75
Tax Reform Act (1969), 75
Tax Reform Act (1971), 75

Taxes, reduced "other," 131, 140–141, 145, 154
Taylor-Woodrow Construction Ltd. (UK), 24
Texas Utilities Company, 60
Third-party power producers, 20–21, 56–57, 78–79, 141–142, 145, 148, 151
"Turn-key" contracts, 58–59

United Kingdom, 24
U.S. Constitution: Commerce Clause, 81
U.S. Department of Agriculture, 16
U.S. Department of Energy (DOE), 16, 22, 24, 70, 83
U.S. Department of the Interior, 17
U.S. Department of the Treasury: IRS, 73
U.S. Windpower (USW), 20–21
Uppland, Sweden, 22
Utility-purchase requirements, 79

Vermont, 138
Vertical-axis wind turbines (VAWTs), 9, 14
VØLUND Company (Denmark), 22

Water and Power Resources Services (WPRS), 17
West Germany, 23
Western Area Power Authority, 141
Westinghouse, 25
Wicks, F.E., et al., 39
Wind-energy conversion systems (WECS), 2, 7, 9
Wind-energy model, 45–47
Wind-energy projects: federal R&D, 9–14; foreign, 21–25; impact with no policy changes, 136–138; major utilities, 18–20; off-shore, 25–28; private, 14–18; third party, 20–21
"Wind-energy property," 84
Wind-Energy Systems Act (1980), 83
Wind parks, 81
Wind project, 93, 100

Index

Wind resource, 93, 94–100
Wind-resource assessment program, 83
Wind turbines, 9–14
Windfarms Limited (WFL), 18–19, 20, 21, 73, 78–79, 81

Winer, B.M., 40
WTG Energy Systems, Inc., 17

Zambrano, T.G., 19

About the Authors

Frederic March, the principal investigator on this project, is an engineering planner and policy analyst who has worked domestically and overseas on issues in energy and environmental policy. While at Arthur D. Little, Inc., he was a member of the Operations Research Section, where he specialized in systems-analysis applications to planning and national-policy problems. Mr. March is now an independent consultant on energy policy and planning.

Edward H. Dlott is currently at the First National Bank of Boston, where he is manager for public finance analysis.

Donald H. Korn is a senior management consultant in the Financial Industries Section of Arthur D. Little, Inc., and has worked extensively in the field of electric-utility investment and incentive policies.

Frederic R. Madio is a mechanical engineer who has specialized in wind energy. He is currently with the Raytheon Corporation.

Robert C. McArthur is a member of the Operations Research Section at Arthur D. Little, Inc., where he specializes in modelling electric-utility systems.

William A. Vachon is a member of the Engineering Science Section at Arthur D. Little, Inc., where he specializes in wind-energy technology.